T0269837

BEES

OF THE WORLD

BEES
OF THE WORLD

A GUIDE TO EVERY FAMILY

Laurence Packer

PRINCETON UNIVERSITY PRESS
PRINCETON AND OXFORD

Published in 2023 by Princeton University Press
41 William Street, Princeton, New Jersey 08540
99 Banbury Road, Oxford OX2 6JX
press.princeton.edu

Library of Congress Control Number 2022938869
ISBN: 978-0-691-22662-0
Ebook ISBN: 978-0-691-24734-2

Printed and bound in Singapore
10 9 8 7 6 5 4 3 2 1

This book was conceived, designed, and produced by
The Bright Press, an imprint of the Quarto Group
The Old Brewery, 6 Blundell Street, London N7 9BH, United Kingdom
www.Quarto.com

Publisher James Evans
Editorial Director Isheeta Mustafi
Managing Editor Jacqui Sayers
Art Director and Cover Design James Lawrence
Senior Editor Sara Harper
Project Manager David Price-Goodfellow
Design Wayne Blades
Illustrations John Woodcock
Picture Research Sharon Dortenzio

Photograph page 2: The face of a *Ctenocolletes smaragdina* female.
The triangular subantennal sclerite that typify the family Stenotritidae
can be seen clearly in this image.

Cover photos: Front cover, left to right, from top: Row 1 Shutterstock/
unpict, /Ed Phillips, Nature PL/John Abbot; Row 2 Nature PL/MYN/
Clay Bolt, Shutterstock/Maksim Vivtsaruk, Nature PL/MYN/Joris van
Alphen; Row 3 Shutterstock/Daniel Prudek, Nature PL/MYN/Clay
Bolt; Row 4 Packer Labs/Sheila Dumesh, Shutterstock/Danut Vieru,
Packer Labs/Sheila Dumesh; Row 5 Nature PL/MYN/Clay Bolt,
MYN/Corne van der Linden, Alamy/Richard Becker.
Spine: Shutterstock/Kletr.

INTRODUCTION

When most people think of bees they imagine an insect about $\frac{3}{8}$ inch (1 centimeter) in length, mostly brown or orange-brown, that lives in enormous, socially complex societies with a single queen and thousands of workers inside a hive. These bees can communicate the location of good food resources through a dance that takes place in the dark but that tells nearby bees which direction to go (in relation to the sun) and how far to travel to find the nectar and pollen they need.

However, most bees are nothing like the one I describe above—the domesticated Western Honey Bee (*Apis mellifera*). This species is native to Africa and southern Europe, but has been domesticated for centuries and transported around the world. It is just one among more than 20,000 species of bee that have been described, almost none of which are anything like it.

In contrast to the Western Honey Bee, the average bee lives a solitary existence in a hole in the ground and will interact with others of its species only for mating, egg-laying (hardly an interaction really!), and perhaps fighting over nest ownership. The second most common lifestyle for bees is nesting in cavities, again primarily solitarily, in plant material such as pithy stems, and in old beetle burrows in wood.

BELOW | A bumble bee *Bombus pascuorum* collects resources from a flower.

The third most common mode of "beeing" is perhaps even less expected: going against the old adage "as busy as a bee," perhaps a fifth of all bee species do no work whatsoever, with the females laying their eggs in the nests of other bees that do the work of nest construction and food collection for them. These are the cuckoo bees. Still other bees make their own nests in abandoned snail shells; on the surface of rocks, stones, or vegetation; inside animal dung; and even in hollow man-made objects such as patio furniture, keyholes, or the fuel lines of crashed aircraft.

ABOVE | The worker honey bee in the center right is performing the waggle dance, while those in the ring around it pay attention. In this way the onlookers find out where good resources may be found.

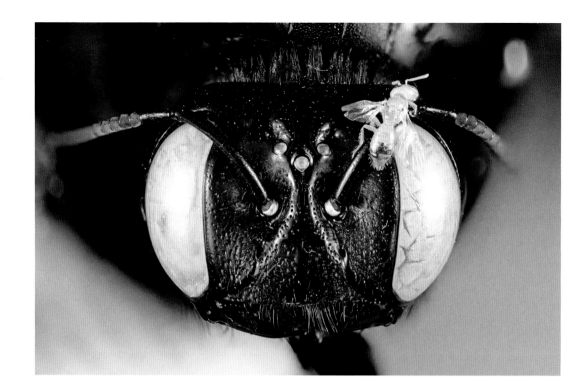

SOCIAL BEES

Like the Western Honey Bee, some other bees live in complex societies. For example, there are at least seven other species of honey bee, one of which is also domesticated in eastern Asia (it is called the Eastern Honey Bee, *Apis cerana*, to differentiate it from the Western species most people know about). Some of these social bees do not nest inside hives, but instead construct their nest on the surface of a tree branch or trunk, on the outside of a building, or on a rock face.

There is also a group of several hundred bee species, restricted to the tropics, called stingless honey bees, which have complex social lives but in ways that differ fundamentally from those of the domesticated honey bees. The bumble bees, comprising several hundred species, are also social. Their societies start off in spring with a single individual that raises a small brood of workers, which in turn help raise more workers—until later in the year, when males and the following year's queens are produced (only a single batch of workers is produced in those species living in colder climates).

Even social bees have other bees that are their natural enemies. Stingless bees suffer from robber bees, which steal both their food and their home (dismantled bit by bit, like taking a neighbor's house down brick by brick to enlarge your own while also raiding their fridge), and the bumble bees include among their ranks socially parasitic bumble bees, in which a single female invades the nest, suppresses the queen (sometimes killing her), and gets the workers to raise her own offspring.

BEE DIVERSITY

Bees are remarkably diverse, not just in nest site choice or social organization, but also in appearance. The smallest bees are less than $1/16$ inches (2 millimeters) in length, and occur in two taxonomic groups: the fairy bees (page 176), which are mostly solitary; and some species among the stingless honey bees, which have complex societies with perhaps 600 individuals living inside a nest the size of a walnut. At the opposite extreme are several bees that can be called giants. If it's the longest bee you're after, then that's *Megachile pluto*, a resin bee that nests inside termite nests and is known from a few islands in the Pacific Ocean.

In contrast, the heaviest bees are probably the queens of some species of bumble bee. Bees also come in a wide variety of shapes, from long and narrow to almost spherical. And they come in all colors of the rainbow, as well as black and white. Many of them do not look like what most people would think of as a bee because they are almost bald and/or have the yellow and black stripes usually associated with wasps. In fact, there are so many bees that don't look like bees and so many other insects that do look like bees that there have been books published on bees that depict other insects on their cover.

So, given all this diversity and confusion, what are bees?

BELOW | A short fat bee *Pachyanthidium* sp. and a long thin bee *Geodiscelis longiceps* demonstrating differences in overall shape among bees.

WHAT ARE BEES?

EVOLUTIONARY ORIGIN OF BEES

To an evolutionary biologist, the simplest answer to this question is that bees are wasps that went down the food chain to collect pollen instead of other animals (or parts of them) as a protein source for their offspring. Bees also collect nectar, not only as an energy source for their own individual activities,

BELOW | A female *Trachusa integra*. The white metasomal scopal hairs can be seen.

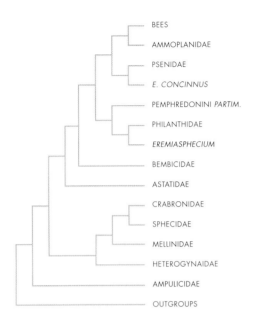

```
┌─── BEES
├─── AMMOPLANIDAE
│ ┌─── PSENIDAE
├─┤
│ └─── E. CONCINNUS
├─── PEMPHREDONINI PARTIM.
│ ┌─── PHILANTHIDAE
├─┤
│ └─── EREMIASPHECIUM
├─── BEMBICIDAE
├─── ASTATIDAE
│ ┌─── CRABRONIDAE
├─┤
│ └─── SPHECIDAE
├─── MELLINIDAE
├─── HETEROGYNAIDAE
├─── AMPULICIDAE
└─── OUTGROUPS
```

ABOVE | A phylogenetic tree of the apoid wasps, showing that the origin of bees is deeply nested among these wasps.

BELOW | Bees have branched hairs, although the form of the branching is diverse—even differing on different parts of the body of a single individual.

but also as a carbohydrate source for their offspring (some bees collect oils for the same purpose). When a yellowjacket (these are not bees, although in some parts of the world they are called meat bees) disturbs your outdoor barbecue or picnic and tries to snatch pieces of hamburger, it is finding a protein source for its juvenile nestmates back home. It might also try to take some jam or other sweet substance from you, which it will use for its own energy needs rather than as food for its relatives.

Bees arose from within a particular group of wasps, tiny little thrips-hunting species, well over 100 million years ago. Clearly it was advantageous to change food sources—there are more than 20,000 extant bee species, yet fewer than 200 wasp species in the group that comprises the closest bee relatives, even though both groups have had the same length of time to diversify.

BEE OR WASP?

However, how do you know whether a particular insect is a bee or not? This is a more difficult question than you might have imagined, knowing now how variable bees are in size and appearance. The answer is not very satisfying: bees have branched hairs somewhere on their bodies, whereas wasps generally do not (but irritatingly, there are a few exceptions). Searching for branched hairs on a bee can take some time, and in some species this defining characteristic is found on a minority of body parts.

scopa

Another characteristic that helps indicate whether a particular insect is a bee is that a particular part of the hind leg is relatively flat and wide, whereas in wasps the same part is usually cylindrical (there are exceptions, however, as in some bees the same part is entirely cylindrical). There are also differences in the structure of the sting apparatus.

None of these characteristics is useful in helping you tell whether an insect flying around in your backyard is, or is not, a bee, and nobody in their right mind would want to cut open the sting apparatus of an insect to find out whether it is a bee or not. There is an easier answer for those patient enough to watch what the insects in their gardens are doing: female bees will actively collect pollen to store on their bodies, usually their hind legs or the underside of their metasoma. However, this strategy fails in the case of bees that take pollen back to the nest inside their digestive system, such as masked bees, and cuckoo bees, which don't collect pollen at all.

Of course, male bees don't collect pollen and are little more than volant sperm donors. Although these can perhaps be identified as bees, because they usually look like slimmer versions of the bees that are collecting pollen and seem to spend most of their time flying around searching for these similar females.

13

BEE ANATOMY

In common with all other insects, the bee body can be divided into head, thorax, and abdomen. But unlike in other insects (other than their waspy relatives), the bee abdomen is not what it seems, because the narrow waist is actually at the junction between the first and second segments of the abdomen. In other words, the ancestor of bees and wasps evolved to have a modified structure so that the first segment of the abdomen is broadly fused to the thorax, and the junction between the first and second segments is very narrow.

BELOW | Side view, of a *Protandrena* female to show the three main body parts with hairs removed. The mouthparts are partially extruded and thus are not accurately in position. The head is rotated somewhat ventrally to the far side.

MAIN PARTS OF A BEE

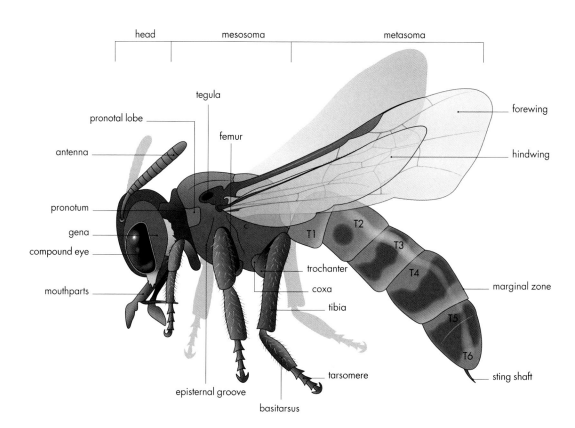

head   mesosoma   metasoma

tegula

pronotal lobe

femur

antenna

forewing

hindwing

pronotum

gena

compound eye

T2

T1

T3

T4

trochanter

coxa

marginal zone

mouthparts

tibia

T5

T6

tarsomere

sting shaft

episternal groove

basitarsus

PARTS OF A BEE'S HEAD

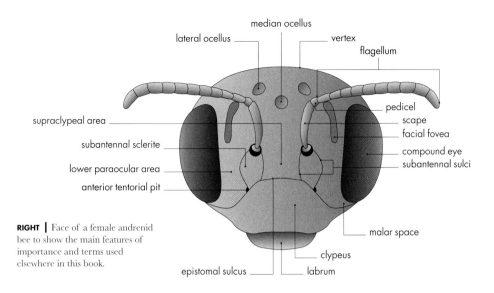

TYPICAL CELLS AND VENATION OF A BEE

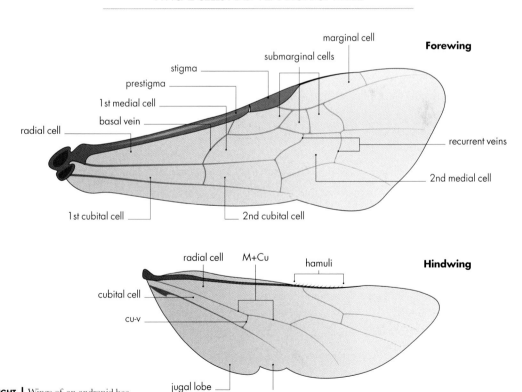

As a result, we need different words for "the thorax plus the first segment of the abdomen" (i.e., the part between the neck and the other narrow part of the body) and "the abdomen minus its first segment." These words are "mesosoma" and "metasoma"—meaning *mid-body* and *hind body*.

HEAD

As in other insects, the bee head has compound eyes, ocelli (three tiny facets that generally measure light intensity), antennae, and the complex suite of features that make up the mouthparts. As discussed below, the details of the structure of the mouthparts are extremely important in bee classification and evolution. The antennae are important in helping us find which sex a bee is: in all but a few cases, there are 13 subdivisions of the antenna in males and 12 in females.

MESOSOMA

The mesosoma bears three pairs of legs and two pairs of wings. Although it often looks as if bees have only one pair of wings, this is because the two wings on each side are held together in flight by a row of hooks, called hamuli. Located on the front edge of the hindwing, these grip into an elongate gutter on the posterior margin of the forewing.

ABOVE LEFT | A female *Andrena labialis*. Although of two families, the adjacent images show two things: the difference in number of antennal "segments" between males and females and the fact males generally have hairier faces than females.

ABOVE | A male *Megachile* showing the longer antennae typically found in males and the modified front legs that are used for the sexually selected, melittological version of "guess who" in some species of the genus.

RIGHT | This *Nomada panzeri* is cleaning its antenna by drawing it through the space formed by the foretibial spur and forebasitarsus.

Each leg is subdivided into five sections: a basal coxa, a small trochanter, an elongate femur, a similarly elongate tibia, and a subsegmented tarsus, the basal segment of which is usually quite long. The tibiae bear one or two apical spurs, that on the foreleg forming part of the antenna-cleaning apparatus. Many bees (often only the females) have a triangular or U-shaped protrusion at the base of the hind tibia—this is the real "bee's knees," in that it helps the insect obtain purchase on the walls of its burrow. Called the basitibial plate, it is generally found in ground-nesting bees and not in those that nest in stems and the like, although here again there are exceptions.

The mesosoma is technically divided into four segments. The first three make up the true thorax and each bears a pair of legs, and the second and third segments bear the wings. As mentioned above, the fourth segment of the mesosoma is the first segment of the true abdomen, called the propodeum. What appears to be the dorsal surface of the propodeum is actually an extension of the third segment of the thorax and is called the metapostnotum in most modern literature on bees. The second segment of the mesosoma is much larger than the others and houses most of the flight muscles. Dorsally, it is subdivided in a way that makes it look as if multiple segments are involved. Fortunately, there are some bees in which these parts are color coded.

BELOW | Dorsal view of the head and mesosoma of an andrenid bee to show terms used frequently in this book.

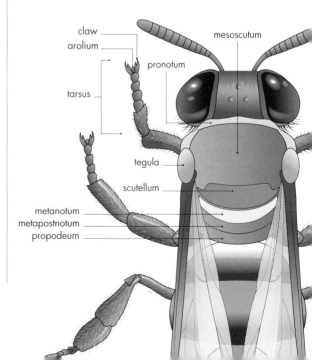

claw
arolium
tarsus
metanotum
metapostnotum
propodeum
mesoscutum
pronotum
tegula
scutellum

METASOMA

The metasoma is divided into segments that, as in the antennae, also differ in number between males and females: females have six and males seven externally visible segments. I say "externally visible" here because both sexes have segments that are telescoped inside the apex of the metasoma, forming the sting apparatus in females and the genitalia in males. These sex-delimited sets of structures are both highly complex and highly variable.

From the sting apparatus, let us consider just the sting shaft—perhaps the best part to choose, because it is the one you can easily see when it pierces your flesh if you have annoyed a bee. In some bees the shaft is curved downward, in some it is curved upward, and in others it is perfectly straight. It also varies considerably in length. Some bees, especially some cuckoo bees, have a sting that is very long and narrow, and that sometimes can even be completely extruded from the body at the end of a structure called the furcula, which might function somewhat like an atlatl (spear-thrower). In other bees, including the stingless honey bees and some different lineages of cuckoo bees, the sting shaft is reduced to practically nothing. Such bees are incapable of inflicting pain with their sting (though the stingless honey bees have other ways of tormenting animals that threaten their nest).

The male genitalia are also remarkably diverse, often even among closely related species. Indeed, it is often necessary to study the male genitalia to be able to identify an individual to species level, because the rest of the body may be indistinguishable among closely related species but the genitalia obviously diagnostic. This is a common feature in insects—indeed, even in animals in general.

BELOW | Ventral view of the genital capsule of a male apid bee.

BEE GENITALIA

gonostylus

penis valves

gonocoxa

spatha, or bridge of penis valves

gonobase

BEE CLASSIFICATION

As usual when experts are involved, bee classification can be a controversial topic. We know that bees arose from among a particular group of wasps, although precisely where in the wasp tree the branch that led to bees diverged is something that not all bee experts yet agree on. There is also disagreement as to how many groups there are at the coarsest level of classification found among bees, which taxonomists call the family level.

In the UK and Brazil, some researchers consider bees to belong to a single family. While other numbers have been considered at one time or another, the most widespread view is that bees make up seven different families. Here, I follow the latter approach, the one advocated by the renowned American melittologist Charles Michener (1918–2015), who spent more than 80 years publishing research articles on bees and had in his prodigious memory a larger proportion of humanity's entire knowledge of bees than anyone else will ever possess.

Some of the controversy over bee classification has arisen from comparing results from traditional morphological analyses with those obtained from DNA sequences. Neither approach is perfect. As most people are not overly interested in the niceties of DNA sequences, it is more illuminating (not to mention aesthetically pleasing) to consider the morphological features that support particular taxonomic groups among the bees. The most fundamental of these groups are the seven families, and to understand how bee experts divide them up we need to pay attention to bee mouthparts and pollen-collecting structures.

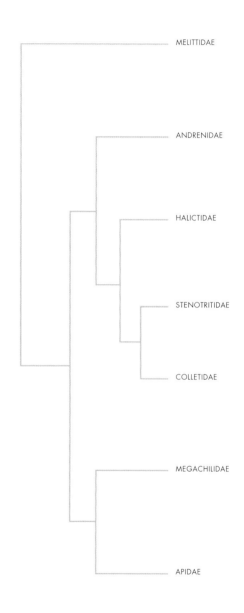

MELITTIDAE

ANDRENIDAE

HALICTIDAE

STENOTRITIDAE

COLLETIDAE

MEGACHILIDAE

APIDAE

ABOVE | The most widely accepted phylogenetic tree for the families of bees.

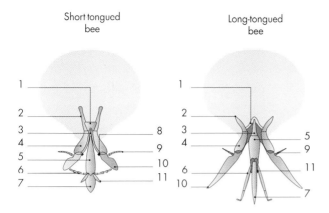

Short tongued
bee

Long-tongued
bee

LEFT | The major difference between short- and long-tongued bee mouthpart morphology is in the structure of the labial palps. Generally, long-tongued bees have two long and two short palpomeres, with the latter at right angles to the former, and short-tongued bees have four equal-sized and equal-oriented palpomeres.

1: lorum, 2: cardo, 3: mentum, 4: stipe, 5: prementum, 6: labial palpus, 7: glossa, 8: lacinia, 9: maxillary palpus, 10: galea, 11: paraglossa.

CLASSIFYING BEES BY TONGUE LENGTH

Living descendants of the group of wasps that bees diverged from more than 100 million years ago fuel their predatory activities through imbibing sugary solutions—most commonly nectar from flowers, but also other sources such as extrafloral nectaries on plants, honeydew from aphids, and so on. But they do not collect nectar as a food for their offspring. Most bees do collect nectar, adding it to the pollen they provide to their offspring: it helps bind the friable pollen into a pollen ball, in addition to providing energy.

Having to collect so much nectar for each offspring places a great burden on the efficiency of the structures that bees use to obtain the sweet solution. So, as soon as the proto-bee arose from the pre-bee wasp, it had to evolve efficient mouthparts for nectar uptake. As a result, bee mouthparts went wild and adaptive radiation provided divergences in structure that were passed on to descendants in each lineage; today, we can study these in order to classify our bees.

A fundamental divergence that must have occurred relatively soon after the origin of bees was the development of the long-tongued bee

morphology, a feature that is shared by two of the seven families (the Apidae and Megachilidae). While this usually does involve a long glossa (the section of the mouthparts that might truly be called a tongue), the real distinguishing feature is the size and orientation of some subdivided, rather leg-like structures (called labial palps) that flank the glossa. In the long-tongued bees there are (with, as always, some exceptions) two very long basal palpomeres, followed by two very short ones that are more or less at right angles to the basal pair. The long ones are flat or concave along their inner margin. In contrast, in the short-tongued bees (the remaining five families) all four palpomeres are usually (there are exceptions here also) similar in length and cylindrical in shape.

LONG-TONGUED BEES

The two long-tongued bee families can be most readily differentiated (in females of nest-building forms at least) by the location of a group of pollen-carrying hairs called the scopa. In the Megachilidae the scopa is primarily on the underside of the metasoma, whereas in Apidae it is primarily on the hind leg. Males and cuckoo forms

of the two families can be told apart perhaps most easily by the shape of the labrum—a flap in front of the mouthparts like an upper lip. In the Megachilidae this is broadest at its very base, whereas in the Apidae it is a little narrower there than it is below. Unfortunately, to observe this it is necessary to open up the mandibles, and these are so strongly closed, especially in Megachilidae, that a beginner (and sometimes even those with considerable experience) may decapitate the specimen in the attempt to identify which family it belongs to.

SHORT-TONGUED BEES

Among the short-tongued bee families, observers must examine a wider range of structures to be reasonably confident of a correct assignment. If the bee's glossa is at least weakly concave at the apex (some have a deeply forked glossa), then the individual belongs to the family Colletidae. An apically concave glossa is a feature these bees share with the wasps, and for a long time many entomologists thought that this suggested the colletids were the first bee family to diverge, with a more pointed tongue being a morphological

ABOVE LEFT | A female mining bee, *Anthophora bimaculata*, on Common Fleabane (*Pulicaria dysenterica*) in the UK, with pollen on the scopa on her hind leg.

ABOVE | The leafcutter bee *Megachile centuncularis* feeding from Common Marigold (*Calendula officinalis*) with pollen on the ventral surface of the metasoma.

RIGHT | Close-up view of the malar areas, labrum, and mandibles of the Himalayan honey bee *Apis laboriosa* to show the narrowing of the labrum towards the base.

feature that originated in the common ancestor of the rest. We are now confident that this was not the case; the concave apex of the colletid glossa is related to a novel use of the tongue.

Colletids are often called cellophane bees because they line their brood cells with a cellophane-like layer (sometimes two or three layers) that acts as a means of waterproofing the brood cell. This material comes from a gland in the bee's metasoma and is applied to the walls of the brood cell with the tongue, which is used like a paintbrush—it clearly wouldn't work as well if it was sharply pointed. This helps prevent the pollen mass from becoming moldy if it is too damp or from desiccating if it is too dry. As it is only the females that work at nest construction and thus need a paintbrush-like tongue, it is perhaps not surprising to find a few species of colletid in which the males have a pointed tongue (and have to be identified to the right family either by using other

structures or, more easily, by otherwise looking like the females of their species).

The closest relatives of the Colletidae are members of the Stenotritidae, the smallest family of bees, with only 21 described species, all of which are restricted to Australia. They are large, very fast-flying bees, with a tongue that is bluntly rounded rather than concave or pointed.

The Andrenidae is a large, diverse family that is found worldwide except for Australia, where they are entirely absent. A defining characteristic of all female and almost all male Andrenidae is the two sulci descending from each antenna, where most bees have just one. The stenotritids also have two subantennal sulci, albeit in a somewhat different form, but as they are only found in Australia and andrenids are found everywhere except Australia, there is no risk of confusing them.

The Halictidae have a truly global distribution (except Antarctica), but a larval melittologist may

have to work hard to be certain of an identification. In most but by no means all halctids, a particular vein—the basal vein—is strongly bent near its base. In most other bees this is fairly straight or evenly curved, or if curved it is not most strongly so at the base. To be absolutely certain of the exceptions, you either have to get to know the bees more personally at lower taxonomic levels, or look for a tiny, usually hairy lobe, called the lacinia. This is usually at the base of a structure called the galea. In halictids it is in the "wrong" position, further toward the base of the mouthparts. So, if the lacinia (which is always hard to find) isn't where it should be, either it's there and you haven't noticed it or you have a halictid (or any of a small group of andrenids, which can be identified by the subantennal sulci, as discussed above).

This leaves us with the second-smallest bee family, the Melittidae, with just over 200 species. To be certain you have a melittid, look at the tongue yet again. If your specimen has the short-tongued bee structure (labial palpomeres all similar in size and shape) *and* a V- or Y-shaped structure known as the lorum with the narrow part toward the tongue base, then it is a melittid. These bees are absent from South America and Australia, so if you are looking at species from either of those continents, you don't need to worry about this family.

TAXONOMY FOR THE BEGINNER

By now you may have given up on becoming a bee taxonomist. Please don't; it is a great deal of fun and, in general, it is easier to learn the overall appearance of a group of bees than it is to identify

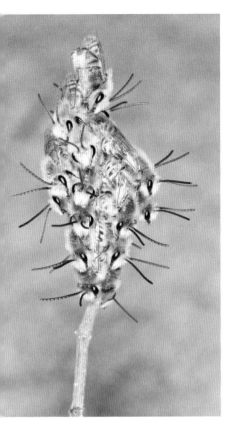

ABOVE | A sleeping aggregation of *Colletes* males.

ABOVE RIGHT | A *Dasypoda hirtipes* female with a full scopal load returns to her nest in the ground.

LEFT | A female *Thyreus waroonensis* searches for the nest entrance of its host.

them by dissecting their mouthparts. Finding out which family a bee belongs to is often the most difficult step in identification in the earlier stages of someone's bee obsession. This is why many geographically restricted identification guides often skip the family level entirely, enabling the user to get straight to the genus level. Indeed, as you look through the generic-level treatments in this book you will find plenty of examples of completely unrelated bees that look, superficially at least, more similar to one another than to their closer relatives. Some of the most striking examples can be found on pages 113, 119, and 220 for very hairy bees, and 109, 126, 154, and 191 for relatively bald ones; some unusually wasp-like bees can be seen on pages 53, 71, 144, and 194.

BEE-NESTING BIOLOGY

Consider the life of an average insect such as a butterfly or grasshopper. After mating, the females wander around looking for suitable places to lay their eggs. For the butterfly this is usually a plant species, one that the caterpillar eats. You can watch cabbage white females flitting to and fro among garden vegetables. After laying a few eggs, they will fly away to lay elsewhere. Most insects take their eggs to where the food is. In contrast, bees bring the food back to where their offspring are (or will soon be): they take pollen and nectar back to the nest. But first, the female has to construct a nest.

NEST SITES

Most bees nest in the ground, with different species preferring different soil types and different kinds of ground cover. Most prefer drier, grainier soils as these are easier to dig into. And most seem to prefer sparsely vegetated ground, perhaps because there's less chance that growing root systems will disrupt their nest. That said, it's easier to find bee nests where there is less vegetation to obscure the entrances, so the proportion that nest in more densely vegetated ground is likely underestimated simply because they are more

RIGHT | Some bees, such as this *Gronoceras felina*, build their nests on the outside of a surface. In this instance a nest made of mud and resin is anchored to the branch of a shrub.

LEFT | A tumulus of sandy soil around the entrance to a *Dasypoda hirtipes* nest.

BELOW | Nests of *Xylocopa californica* revealed in sotol (*Dasylirion* sp.) stalks that have been split down the middle. The brood cells are separated by disks of chewed sawdust.

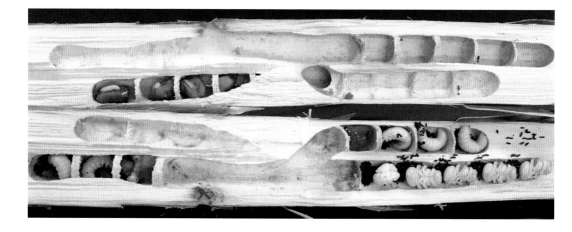

through quite hard wood, and they and other species with similar nest site preferences can be pests in wooden buildings.

Other hollow structures can also be used as nests, ranging from snail shells to cavities in brick walls. There have been cases of bees plugging up stethoscopes in a field hospital or rendering keyholes unusable. However, claims that bees were the cause of a plane crash through plugging up fuel lines have been shown to be false, with nesting beginning after the crash occurred: bees not guilty.

NEST FORMS AND CONSTRUCTION

Relatively few bees construct a nest on a surface by molding resin, mud, or a ball of plant hairs into a structure for brood cells. Some will decorate such nests with gravel and make a mosaic. Even rarer substrates that bees make nests in include dry dung, plant galls, termite nests, and rodent burrows (particularly popular among some bumble bees).

Bee nests in the ground are usually not simple tunnels. Inside each nest a female will construct brood cells, one for each offspring. Some species build these adjacent to the main tunnel, while others dig side branches with one cell at the end of each. Others build the cells in clusters, sometimes constructing a cavity around them, while others make a row of brood cells connected end to end. Some bees make shallow nests, sometimes just beneath the soil surface. The deepest nests ever found were more than 16 feet (5 meters) deep.

The brood cells are where the bee offspring develop, and they have to be constructed with care. They are usually lined with waterproofing materials, most commonly waxy glandular secretions that make the inside of the brood cell look shiny. These serve both to keep excess water out of the brood cell and to maintain appropriate levels of humidity inside, so that the food supply does not go moldy or dry out. In a few species the

difficult to find. Despite these overall preferences, some bees preferentially nest in densely grassed lawns and may be pests of golf courses. Some bees prefer to nest on slopes, others on flat ground, and still others in vertical banks. The insects often favor south-facing slopes (in the northern hemisphere, or north-facing ones in the southern hemisphere) because of the increased warming effects of the sun. That said, some cool-adapted bees can persist in warmer climates by nesting on the shadier side of a hill. To a bee, the sandy mortar between bricks in a wall, or dried-mud adobe, is like a cliff and can provide suitable opportunities for nesting, with some species preferring such substrates.

Pithy stems can be relatively easy to dig into, and many bees use dry raspberry canes or other stems to nest in. Other bees will be able to use a stem only if something else has hollowed it out first. The harder the wood, the more effort it takes to chew a burrow to make a nest, so lots of bees that nest in harder wood rely on other insects (often beetles) to do the hard chewing for them. Some large carpenter bees can chew their way brood cell lining is unusually thick and serves as an additional food source for the larvae. Cellophane bees make a polyester brood cell lining, and so the offspring literally develop inside a plastic bag. The provisions their mother gathers are often relatively liquid made possible by the lining's greater leak-proofing properties. Other bees line their brood cells (and often the entire burrow) with extraneous materials: pieces of leaves or petals, resin, and even pieces of plastic bag or kitchen tile caulking have been used in some cases. Attentive wool carder bee mothers shave hairs from plant leaves and construct a large fluffy ball, which they hollow out before collecting food for their offspring. You have to admire a mother that makes a pillow for her offspring to grow up inside as well as collecting all the food it needs.

LEFT | An opened brood cell of a *Megachile*. To the left is the pollen and nectar mixture that has yet to be consumed by the larva, which is to the right. Note the leaf sections that surround the contents of the brood cell.

BELOW LEFT | A brood cell cluster of a *Bombus pascuorum*. The large queen is to the left and a recently emerged worker is at the bottom.

BELOW | A female *Anthidium manicatum* forming a ball from the hairs she has shaved from the surface of a leaf.

THE BEE
LIFE CYCLE

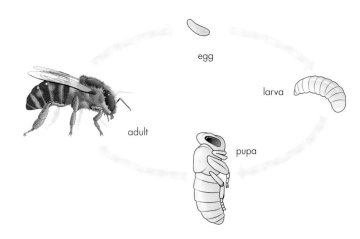

egg

larva

adult

pupa

Bees start life as an egg, usually banana-shaped, laid on top of a pollen ball. Cuckoo bees are more diverse in egg structure, as they have to place the egg in a way that reduces the chance of detection by the host bee. As a result, cuckoo bee eggs are often flattened on one side or at one end and flush with the brood cell surface, with the rest of the egg buried in the brood cell wall.

The bee mother determines which sex her offspring will be: if she passes stored sperm over the egg as it passes through her reproductive tract, then it will be a female (with rare exceptions); if she does not, it will become a male. In other words, male bees result from unfertilized eggs and females from fertilized ones. This unusual form of sex determination is also found in wasps, ants, sawflies, thrips, and some other arthropods.

As a result of their rather easy lifestyle inside a brood cell, with all the food they will ever need, bee larvae are morphologically rather simple "couch potatoes," especially in comparison to the elaborate forms found among caterpillars, which have to fend for themselves "outdoors." Like all juvenile insects, bee larvae have to shed their exoskeleton (which includes the lining of the gas-exchange tubes that run through their bodies and much of the lining of their digestive system) in order to grow, because their exoskeleton is rather inflexible and cannot stretch much. This is akin to

needing only five sets of clothes from birth to adulthood, as bee larvae must "change their skins" a number of times between hatching from the egg and becoming fully grown. Then the larva molts to the pupal stage.

Bee pupae have the same overall structure as the adult, except that the wings are reduced to little flaps and the legs and antennae are held close to the body. Pupae start out the same white color as the larvae, but gradually darken as the integument of the adult forms beneath the translucent pupal skin. Upon emerging from the pupa, the wings expand as blood is pumped into them. Once the exoskeleton has hardened and the wings have dried, the adult is ready to leave its natal nest and fly.

The timing of the life cycle with the seasons varies greatly among bees, with different species being active as adults at different times of year. Many of the earliest spring bees have overwintered as adults and will have completed their brood production by the time the season's willow catkins have faded. Others will pass the cold season as fully grown larvae, and then pupate and develop into adults once the weather warms. Desert bees survive—sometimes for years—as fully fed larvae, ready to pupate as soon as soil humidity levels suggest that it has rained enough for flowers to start developing.

BEE FOOD AND POLLINATION

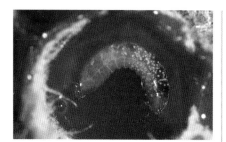

ABOVE | This *Scaptotrigona depilis* larva is preferentially feeding on *Zygosaccharomyces* fungi growing on its brood-cell lining rather than on the pollen and nectar in the brood cell.

BELOW | This *Agapostemon* female is using sonication or "buzz pollination" on a pale meadow beauty blossom. The pollen grains can be seen like shooting stars.

Most bees are mass provisioners, meaning that all the food required for the development of the larva is collected before the egg is laid. This is like collecting 18 years' worth of groceries, piling them inside a room, giving birth, and then leaving your child to develop alone. In most cases, the food is pollen and nectar, although some bees replace nectar with floral oils. However, it has recently been suggested that much of the nutrition most developing bees get does not come directly from the floral sources collected by the mother, but from microbes that grow on the pollen and nectar mixture. In other words, much of the nutrition the bee larva gets is probiotic. Certainly some bees that make a very liquidy provision mass have brood cells that smell "yeasty," indicating that fermentation has taken place. Some bee larvae have even been shown to preferentially consume the fungi growing around the edges of their brood cells.

ABOVE | As the genus name suggests, *Nolanomelissa toroi* females collect pollen only from flowers of plants in the genus *Nolana*.

With the enormous diversity of flowers available to any mother, how does she choose which ones to visit? Some species are oligolectic, which means they will collect pollen from only one, or a few closely related, plant species. An example of such a species is the sundrop sweat bee (*Lasioglossum oenotherae*), which like its closest relatives collects pollen only from evening primroses, including the horticultural sundrops that many people in eastern North American cities have in their gardens. These are flowers that open at night or under the low light intensities associated with dawn or dusk. Thus, the sundrop sweat bee is active in the early hours of the morning and goes to bed "for the night" at 9 a.m.! At the opposite end of the spectrum are bees that will visit almost any flowering plant for resources, our domesticated Western Honey Bee coming close to that. Most bees fall between these extremes, and it is normal for the insects to visit a wider range of flower species for nectar than for pollen.

While most nest-building, food-collecting bees have developed special hairs to carry pollen—the scopa, usually on the hind leg or the underside of the metasoma—others have evolved mechanisms to obtain pollen in less usual ways. A particularly interesting example is the genus *Samba* (page 48), where the bee uses the mandibles, midlegs, and hind legs simultaneously to obtain pollen. Likely the most common additional adaptation for collecting pollen is buzz pollination.

Many flowers have pollen on the inside of their anthers rather than easily accessible on the outside, so it has to be shaken out of the anthers if the bee is to get it. Buzzing bees seem preadapted to do this: they land on the flower, hold it tightly with their mandibles and/or legs, and vibrate their flight muscles. The result is usually a fairly high-pitched buzz, and the pollen pours out of the anther and collects on the hairs on the bee's body. Blueberries and other ericaceous plants require buzz pollination, as do Tomato plants (*Solanum lycopersicum*) and a wide range of wildflowers.

As some bees visit only one or a few species of flowers for pollen, do the plants rely on the insects for pollination? It is generally true that bees rely on particular flowering plant species more than particular flowers need the bees, thus, with exceptions (such as some cacti where most pollination results from visits of a single bee species), it is a somewhat one-sided relationship.

SOCIAL BEES

LEFT | Most stingless bees have beautiful nest entrances. In this instance *Pariotriogona klossi*—a tear-drinking bee—has built a nest on a limestone cliff, and the nest has numerous tubular entrances that lead to a conduit and thence a fissure in the cliff wall.

RIGHT | Bumble bee colonies are initiated by a foundress in spring. She produces a brood of workers that labor to increase the colony size before there is a switch to the production of males and the next generation of reproductive females. This drawing shows a species with a long colony cycle; some species produce males and next year's queens in mid-summer.

As discussed at the start of the chapter, most bees are solitary. However, several hundred species are incapable of living outside of a large perennial colony headed by a queen that survives for several years. Honey bees and stingless honey bees fit that description, as does *Lasioglossum marginatum*, a very unusual sweat bee with queens that live five or six years—longer even than honey bee queens (pages 106–7).

Other species have annual colonies, started in spring by a single overwintered female—the foundress—that mated late the previous year and has stored sperm so she can fertilize eggs after emerging from diapause (dormancy). Bumble bees (page 206) are the best-known examples of species that form annual colonies, which can get quite large, with hundreds of workers. There are also lots of sweat bees with annual colony cycles, although most of them are much smaller with perhaps half a dozen workers at most. There are even examples where the average number of workers is less than one—meaning that some females in the population are entirely solitary, whereas others have just one or two workers.

The types of social insects we are most familiar with—in addition to the bees mentioned above, yellowjackets, hornets, and ants—are all eusocial, meaning that normally the workers are the daughters of the queen and help her raise the next generation. But there are other forms of social life among the bees. A few bee species have workers that are sisters to the reproductively dominant queen in what is termed a semisocial society. This is rare when it is the only form of social organization, but more common as a phase during the cycle of a colony that is eusocial at some point.

It can arise in two ways: either as a grouping of sisters in spring, in which the dominant one remains as a queen when the brood develops and the society becomes eusocial; or if the queen in a eusocial society dies and is replaced by one of her daughters. In the latter case, usually the oldest or the largest daughter becomes the replacement queen and her sisters collect the food on which she then lays her eggs.

Some bees share a nest but once underground act as if they are solitary. In these communal societies the bees share a nest entrance but have their own branches below ground, construct their own brood cells, collect their own pollen and nectar, and lay their own eggs. This is more like an apartment building style of living, as opposed to an extended single-family dwelling.

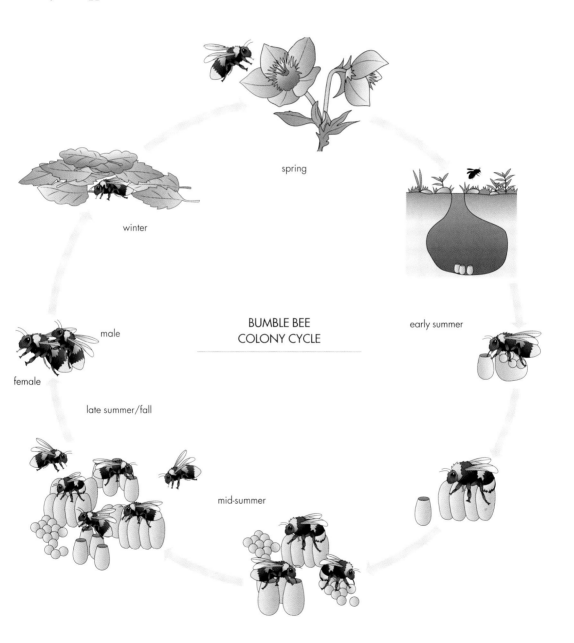

spring

winter

early summer

male

female

late summer/fall

BUMBLE BEE
COLONY CYCLE

mid-summer

ENEMIES

CUCKOO BEES

Bees make large amounts of nutritious food available for their offspring, but those resources can be stolen by other animals. Prime among these thieves are other bees known as cuckoo and socially parasitic bees. Bees seem to have evolved cuckoo bee behavior on many occasions, and while the precise number remains unknown (as new cuckoo bee lineages are discovered every now and then), such larceny has arisen at least 19 times during bee evolution. Some origins involve few species that are very closely related to their hosts, while others involve hundreds. One group of the family Apidae, the Nomadinae, contains more than 1,600 species of mostly wasp-like bees that lay eggs on provisions provided by hosts from five of the seven bee families, including other apids.

ABOVE | This *Sphecodes* female is leaving its host nest, perhaps after having laid an egg on a pollen ball constructed by the host.

RIGHT | This female *Philanthus triangulum* wasp has paralyzed a honey bee worker, which it will place in a brood cell along with others before laying an egg. The wasp larva will then eat the paralyzed bees.

Cuckoo bees have a variety of ways of entering a host nest. Most enter and lay an egg while the host female is out foraging, while others break in after the nest is completed. The latter must have structures that permit a break-in—for example, if the host is a cellophane bee, then the cuckoo has to cut through the plastic and often has saw- or forceps-like structures at the tail end that facilitate this.

While some cuckoo bee females eat the egg they find in a completed brood cell, most leave it to their offspring to make sure the host offspring is dispatched: early-instar cuckoo bee larvae often have long sickle-shaped mandibles with which they slice and dice the host egg or early-stage larva (and other cuckoo bee early stages if the brood cell has been attacked more than once). Some particularly enterprising cuckoo bee mothers lay their egg directly on the egg of the host, and the newly hatched cuckoo larva (which develops more quickly than the host egg) then consumes the contents of the host egg before consuming the provisions.

Bees that lay eggs in the nests of social bees have more fearsome obstacles to overcome. Some do so by brute force, with heavily armored bodies that can withstand the attacks of the workers as they attempt to defend their colony and strong mandibles and/or stings that can maim or kill the defenders. Others use stealth. Some cuckoos of communal bees have relatively flat bodies that make them less noticeable among the thrum of activity as they go up and down the brood tunnels. Others have evolved to become social parasites that take over the society they invade.

OTHER ENEMIES

It's not just other bees that are natural enemies of bees: all
sorts of different animal groups have evolved to obtain food
from bee nests or even adult bees. Bee wolves are wasps that
are not too distantly related to the bees themselves. They
include dozens of species and are found almost worldwide,
and one species concentrates on attacking honey bees. These
wasps provide paralyzed bee adults as food for their offspring.

There are also flies that attack adult bees, including
thick-headed flies, which lay their hooked eggs into the
bodies of their host. The fly larvae consume the insides
of their hapless host and pupate in the hollowed-out shell
that was once a healthy, active bee. Various predatory flies
will include bees in their diets, although few specialize on
bees. Similarly, assassin bugs and their relatives are as
happy sucking the juices from a bee as they are from any
other insect, although there are some that specialize on

ABOVE | This *Physocephala*,
a big-headed fly, is on the lookout
for a bumble bee, which it will
pounce upon to oviposit. The fly
larva will feed on the insides of
the bee, eventually killing it.

RIGHT | This *Diadasia* male has
inadvertently picked up some
triungulin larvae of a meloid beetle,
which will be transferred to a female
when he mates. The beetle larvae
will be taken back to the nest by the
female, where they will feed on
the developing bee larva and its
pollen ball.

attacking social bees—skulking outside the entrances to their nest and picking off foragers as they leave or return.

Numerous other insects consume bee broods rather than the adults. Many species of bee fly (so called because they look like bees, not because they attack them) flick their eggs down the entrances of bee burrows. The larvae that hatch then find their way into a host brood cell and hide until the host larva is fully grown. At this point the bee fly larva sinks its mandibles into the host, benefiting from the larger meal. Some groups of beetles have found ways of getting their active young larvae inside bee nests. These small juveniles latch onto a bee in the hope that it's a female that will take it back to the bee nest, where it then leaves the host and consumes the bee food or brood.

Large predators have difficulty getting at the broods of bees that nest underground, but eating the contents of nests in stems or wood is easier. Woodpeckers can chip away at the wood around a carpenter bee nest, for example, and smaller insectivorous birds can peck holes in stems to get at smaller pith-nesting bees. Such bees are also easier hosts for parasitic wasps that lay eggs on or inside a developing larva.

Most bee enemies are rather specialized in their tactics and attack a small proportion of bee species. There is, however, one species that seems to have a negative impact on most bees, and it is the one the writer and readers belong to. So how can we counteract our negative effects?

HOW TO HELP BEES

A major concern about our food supply is the impact of colony collapse disorder on Western Honey Bee populations. As this is an introduced species in most parts of the world, conflating helping domesticated bees with saving wild bees is erroneous at best and disastrous at worst: Western Honey Bees, at least under certain circumstances, can deplenish the resources other bees need, and augmenting their populations likely means reducing those of native species.

Climate change seems to be causing the geographic ranges of some bee species to become restricted. Bumble bees seem to be losing their southern ranges (in the northern hemisphere, where most bumble bee species are found) but not compensating by expanding northward. Bees living on mountaintops will have nowhere to go if their limited habitat warms up too much for them. Pesticides are as much a problem for wild bees as they are for domesticated honey bees, and have been shown to be seriously involved with colony collapse disorder in Western Honey Bees and mass die-offs of bumble bees.

Habitat loss might be more problematic for bees than most other organisms for several reasons, including the fact that they reproduce in a nest and need resources in good supply within their foraging range. Diseases cause problems for wild bees, especially as it is becoming increasingly clear that they can be transmitted from domesticated bees (honey bees and bumble bees in particular) to wild bees. Introduced species are also a serious threat, in some cases through transmitting novel diseases—as has happened with both North and South American bumble bees as a result of intercontinental movement of species. But there are also threats arising from competition for resources, sometimes including direct aggression: *Megachile sculpturalis*, introduced to North America, has been seen gumming up the bodies of large carpenter bees as part of its takeover of the latter species' nests.

There are many things you can do at home to help bees. If you have a garden, avoid applying chemicals, use as little mulch as possible, and leave old pithy stems and wooden structures in situ for bees to nest in. Even if you live in a high-rise and have only a planter on a small balcony, bees may nest in the soil and forage from the flowers. This will be most useful if other building residents act similarly. "Bee hotels" are artificial constructs,

LEFT | Dead honey bees in the hands of a beekeeper due to colony collapse disorder in Bavaria, Germany.

RIGHT | *Bombus affinis* is listed as an endangered species in both the USA and Canada.

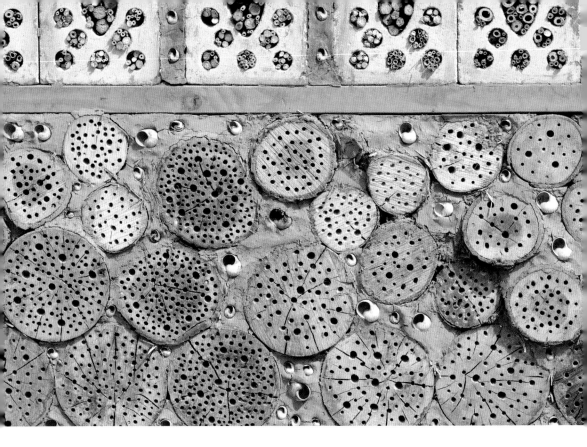

the production of which has become a cottage industry in some parts of the world. They need to be used with care: they are sometimes more attractive to non-native bee species or wasps (wasps are fine by me, and they are often good biological control agents of pests in the garden), and some designs are likely to cause deaths of bees and/or their broods. Some designs are beautiful works of art.

Even without access to a patio or a backyard you can help bees by buying organic food if you can afford it. Encouraging local and larger-scale governments to use bee-friendly policies can have a large impact. You can also become a citizen scientist by taking pictures of bees and posting them on online platforms such as iNaturalist (inaturalist.org) and Bumble Bee Watch (bumblebeewatch.org). While this doesn't help the bees directly, it does help scientists find out more

about the distribution of bees and the way this is changing. Of course, to be a good scientist it helps to know what bees look like. Perhaps one of the biggest threats to wild bees is that so few people know what they look like or what they need to survive. This book aims to bring this enormous diversity of appearances to the front of your mind: getting to know bees should be as much an aesthetic experience as it is an environmental or scientific one.

ABOVE | Bee hotels are becoming increasingly common as people try to "save the bees." It remains to be determined whether so many potential nest sites in a small area will end up attracting too many natural enemies, such that their effect on bee populations is the opposite of that desired.

ABOUT THIS BOOK

This book treats 104 of the 500-plus genera of bees.
Choice of genera to include was based on a combination
of those that have images available and those that helped
illustrate the diversity of bee appearance and biology.
An attempt was made to include all bee tribes, although
while there is no broad consensus as to the delimitation of
bees at the tribal level of classification (which falls between
subfamily and genus in terms of the levels that are in
common usage), images of some were not available.

ABOVE | Male *Osmia bicornis* bees
emerging from their "hotel."

The following accounts are arranged according to the phylogenetic interrelationships among the bees. This is true both for the seven families and, within each family, for the subfamilies and, where appropriate, for the tribes also. The sequence for families will seem unfamiliar to those used to treatments based upon earlier phylogenies (as in Michener 2007) or arranged alphabetically. The subfamily classification is also perhaps unfamiliar to some readers—although it is based, as far as possible, on the most recent phylogenetic understanding of the relevant group of bees. Not all melittologists will agree with these recent findings.

Within the tribe (or subfamily if there are no widely accepted tribes), bees are treated alphabetically, for two reasons. First, many genera have not been included in recent phylogenetic studies of higher-level bee

THE BEE FAMILIES

groups. Second, generic-level phylogenies are those most likely to be unstable and susceptible to change as a result of additional research.

Each generic treatment contains a discussion of the biology of the genus and one or more images. Most images are of live bees, but some particularly interesting bees are rare enough that only images of specimens from collections are available.

At the end of each generic treatment there are three additional sections of text and a map. The "distribution" text describes the geographic range of the genus and broadly matches the information provided on the map. However, the maps are, by necessity, somewhat coarse and it is quite likely that additional records outside of the ranges shown are known to some melittologists. The "habitat" section describes the kinds of areas within the geographic range where species in the genus might be expected to be found. The "characteristics" section lists some of the features that help to identify the genus, although note that these are not always sufficient to confirm an identification as a complete list would often be very long and technical.

LEFT | While honey bee workers are well known, the drones are much less often found and may look surprisingly unfamiliar.

RIGHT | This *Andrena haemorrhoa* female has the hind tibia entirely covered in scopal hairs, unlike the tibia of a honey bee worker which is largely bald on the anterior surface.

MELITTIDAE

With 14 genera and just over 200 species, this is the second-smallest bee family. Melittidae are found only in North America, Europe, Asia, and Africa. They are short-tongued bees with a pointed glossa, a proboscideal tube, and a fully exposed mid-coxa. Many resemble *Andrena* (page 58), but differ in the last two of the above characteristics and in having only one subantennal sulcus and lacking facial foveae. All Melittidae nest in the ground and most are oligolectic.

Despite the small number of species, melittids are morphologically diverse with three subfamilies. The Dasypodainae have a largely bare or absent paraglossa. It contains seven genera: four restricted to Africa, two extending into the Palearctic, and one restricted to North America. The Meganomiinae and Melittinae have relatively large, hairy paraglossae. Meganomiines are distinguished by lacking an apical mandibular tooth in females and the apex of the marginal cell is rounded and bent away from the wing margin. There are four genera and 11 species, mostly with extensive yellow markings. Melittines have at least one subapical mandibular tooth and a marginal cell apex that is at least close to the anterior wing margin. They are restricted to Africa, Madagascar, and the Arabian Peninsula, and include three genera, one in Africa, one throughout the northern hemisphere, and one in the northern hemisphere and Africa.

DASYPODA

Females of this eastern hemisphere bee genus of 39 species have very long scopal hairs on the hind legs, making them look as if they are wearing baggy pants. Their generic name reflects this, as it means "hairy feet." The species illustrated here has the species name *hirtipes*—which also means "hairy feet" (another name for the same species was *Dasypoda plumipes* which means "hairy feet, feathery feet"!).

Dasypoda nest in sandy soil, often making large tumuli from the excavated soil at the nest entrance. One species holds the record for the largest nesting aggregation ever discovered, with more than 12 million burrows estimated along a riverbank in Russia. Some species have adapted to urban living by nesting between the cobbles in cobblestone streets, and keep on provisioning the brood even when a car has run over the tumulus

GENUS
Dasypoda

DISTRIBUTION
Mostly temperate Eurasia, from the UK to Japan, and northern Africa as far south as Ethiopia

HABITAT
Semiarid areas, meadows, fields, and open woodlands, wherever suitable sandy soil and foodplants occur

CHARACTERISTICS
- Lacking pale integumental markings
- Two submarginal cells
- Female mandible toothed
- Hind tibia lacking keirotrichiate area
- Paraglossa small and slender, largely lacking hairs
- Basitibial plate absent

ABOVE LEFT | A *Dasypoda hirtipes* female demonstrating the hairy hind legs that give the genus its name.

ABOVE | Even male *Dasypoda hirtipes* have hairier legs than the females of some bees.

at their nest entrance. Their brood cells are not lined with any waterproofing but their pollen balls are unusually shaped, with conical "legs" or ridges on the underside reducing contact of the food with the brood-cell floor.

SAMBA

This genus of only 11 species of ground-nesting bees comprises the entire tribe Sambini. In earlier treatments it was divided into two genera, *Haplomelitta* and *Samba*, but as it seems the latter arose from within the former, and *Samba* is the older name, they are now united as one genus with multiple subgenera.

Females of *Samba* subgenus *Samba* have an enormous sickle-shaped hind tibial spur, the function of which was unknown until Kenyan entomologist Dino Martins observed females raking the anthers of *Crotalaria* flowers between the spur and the hind basitarsus. To open the flowers sufficiently to access the anthers, the bee has to use both mandibles and midlegs. This seems to be a unique pollen-collecting mechanism among the bees.

LEFT | A female *Samba turkana* about to alight on a flower of its host, *Crotalaria*.

GENUS
Samba

DISTRIBUTION
Southwestern and eastern Africa

HABITAT
Semiarid scrub to woodland openings where the host plants are common

CHARACTERISTICS
- Two submarginal cells
- Vertex straight to concave
- Paraglossa very small to absent
- Mandible of female with subapical tooth
- Male gonostylus clearly separated from gonocoxa
- Male clypeus same color as rest of face

MELITTA

There are 51 species in this genus, which is found in most of the northern hemisphere (although it is absent from central USA), as well as eastern and southern Africa. *Melitta* are usually nondescript bees, superficially resembling a cross between a honey bee and a large *Andrena*. Some species even have orange markings on the metasoma, increasing their similarity to honey bees.

These are solitary bees that nest in the ground and are oligolectic, each with one or a few closely related floral hosts. One eastern North American species specializes on the pollen of cranberries and related plants. A European species that visits Harebells (*Campanula rotundifolia*) has the scientific name *Melitta haemorhoidalis*—in honor of its red tail.

RIGHT | This *Melitta* male looks like an archetypal solitary bee. However, as you can see from the other images in this book, this standard view ignores the spectacular diversity among the bees.

GENUS
Melitta

DISTRIBUTION
Western and eastern North America, southern and eastern Africa, and Eurasia

HABITAT
Diverse, from arid and semiarid areas to moist meadows and open woodland, wherever suitable foodplants occur

CHARACTERISTICS
- Lacking yellow or white integumental markings
- Paraglossa densely hairy
- Mandible of female with subapical tooth
- Three submarginal cells
- Metapostnotum large and dull

REDIVIVA

These are long-legged bees; females of most species have elongate forelegs, some remarkably so. The long legs serve to extract floral oils from the end of floral spurs of their *Diascia* hosts (the bees are less restricted in choice of pollen sources). The ends of the legs have especially dense hairs to absorb the oils, which are used instead of nectar as an energy source for the offspring. The oils are transported back to the nest among the scopal hairs of the hind legs. There has been coevolution between the depth of the floral spurs and the length of the bees' foreleg. The bees fly with their legs folded, but upon landing on the flower they are immediately thrust into the floral spurs. There are 33 species; one group of seven have reverted to nectar collection from oil-collecting ancestors.

BELOW | This *Rediviva steineri* has quite long front legs, but there are other species in the genus with even longer legs than this one.

GENUS
Rediviva

DISTRIBUTION
Southern Africa

HABITAT
Arid and semiarid areas where their specific foodplants occur abundantly

CHARACTERISTICS
- Lacking yellow or white integumental markings
- Mandible of female with subapical tooth
- Three submarginal cells
- Metapostnotum small and shiny
- Paraglossa densely hairy

This is a genus of 16 species of oil-collecting bees, replacing nectar with floral oils as an energy source for their offspring. All species collect pollen and oil from *Lysimachia* flowers (commonly called yellow flag or loosestrife). They obtain the oil with their tarsi and transport it back to the nest in the scopa on their hind tibiae and basitarsi. The oil is also used as waterproofing to line the brood cells. As their hosts produce no nectar, the adult bees get the energy they need from nectar from other flowers. Some horticultural varieties of *Lysimachia* produce neither oil nor nectar, so gardening to attract *Macropis* bees is not straightforward. Nests are in the ground, often in quite damp soil.

BELOW | A *Macropis* female on its Yellow Loosestrife (*Lysimachia punctata*) host, from which it gets both pollen and floral oils, but not nectar.

GENUS
Macropis

DISTRIBUTION
Throughout the northern hemisphere except the driest and coldest areas

HABITAT
From cold temperate regions to the humid tropics wherever foodplants occur in abundance

CHARACTERISTICS
- Lacking pale integumental markings, except on the face of males
- Paraglossa densely hairy
- Two submarginal cells
- Male with well-developed pygidial plate

As the prefix "Mega" suggests, these are large bees, often more than ¾ inches (2 centimeters) in length. There are five species, four of which have extensive yellow markings, making them appear like large, fat wasps. The fifth is all black and the single specimen known was misclassified (as an *Andrena*) for more than a hundred years, until its true affinities were recognized in 2019. The genus is found from southern Africa to Somalia and also on the Arabian Peninsula. The males have a large number of features that seem to be attractive to females during courtship and mating, including modified legs, antennae, and underside of the mesosoma. The males are unique among the bees in producing stridulatory vibrations during mating. *Meganomia* make deep nests in the ground and their larvae spin cocoons.

LEFT | The flattened apical parts of the antenna can be seen clearly in this close-up of the face of a male *Meganomia binghami*.

RIGHT | A female *Meganomia gigas* demonstrating its distinctly wasp-like color pattern.

GENUS
Meganomia

DISTRIBUTION
South Africa to Somalia and Yemen

HABITAT
Arid to subarid areas

CHARACTERISTICS
- Large, bulky Melittidae, ⅗–⅞ in. (15–22 mm) in length
- Mandible of female without subapical tooth
- Usually (four of the five species) with extensive yellowish markings on head, mesosoma, and metasoma
- Arolia absent
- Male antenna with curled and flattened apex

When a bee looks more like a wasp, as is the case with some *Meganomia*, this is often the result of mimetic resemblance: once a predator has been stung or otherwise injured by an organism with a particular color pattern, it will tend to avoid other potential prey that look similar. In the case of a female bee that stings, this is called Mullerian mimicry—two or more species have a similar color pattern because a predator only has to be stung by one of them to learn to avoid both. But what of the males? They cannot sting because the sting is a modified ovipositor.

ANDRENIDAE

Members of the family Andrenidae are commonly referred
to as solitary mining bees. While it is true that they all nest
in the ground and most of them are solitary (some are
communal, but none have queens and workers), mining bees
that are solitary are found in all seven bee families, so this
common name is not very useful. A defining characteristic of
Andrenidae is the two sulci leading from each antennal socket
and reaching the clypeus. This sets off a small area beneath
each antenna called the subantennal sclerite. Some other
bees also have two subantennal sulci, but except for a few
examples where the second sulcus is difficult to find, such bees
are restricted to Australia (family Stenotritidae; see page 111
and the photograph on page 2). As there are no Andrenidae
in Australia (they are found on all continents except Australia
and Antarctica), there is little chance to confuse the two
families. Alas, there are some Andrenidae in which one of the
subantennal sulci has been lost and others where it is hard to
see among the coarse sculpture of the face.

sulci

sulcus

The family Andrenidae consists of three subfamilies: Andreninae, Panurginae, and Oxaeinae. Andreninae includes six genera, but just one of these contains more than 99 percent of the total of 1,564 described species, while the remaining genera combined have only 11 species. Andreninae species are found throughout the northern hemisphere, with relatively few in Africa and tropical Asia, and only four in South America. Panurginae also contains many species (more than 1,430) but has many genera (at least 30). They are found throughout the northern hemisphere, but with the greatest diversity in South America and around the Mediterranean. Oxaeinae has a small number of genera (four) and a small number of species (22). It has a primarily South and Central American distribution, with a few species in the southern USA.

The three subfamilies can be separated by features of the stigma and marginal cell: the former is almost absent and the latter very long and narrow in the Oxaeinae, whereas the other two have a relatively large stigma and the marginal cell has more normal dimensions. The Andreninae can be differentiated from the Panurginae by the placement of the scopa: this is primarily on the hind tibia in Panurginae but on the hind femur and tibia as well as the propodeum in most Andreninae (see page 12). Most, but not all, Panurginae moisten the pollen with nectar while it is still on their hind legs.

Commonly referred to as solitary mining bees and with more than 1,500 species, *Andrena* is more speciose than any other bee genus bar one—*Lasioglossum* (page 106). Given the large number of species, it is not surprising that they are very variable in superficial appearance. They include small black species, mid-sized species whose metasoma has bands of white hairs, and large species that may have a bright orange integument

OPPOSITE | The beautiful *Andrena fulva* can be a pest of lawns but makes up for that by pollinating fruit trees.

BELOW | A female *Andrena vaga* takes a rest before starting to collect pollen from willow catkins.

GENUS
Andrena

DISTRIBUTION
Almost worldwide except for South America and Australasia

HABITAT
From temperate woodland to semiarid scrub habitats, and from sea level to high altitudes

CHARACTERISTICS
- Stigma present and distinct
- Facial fovea of female with velvety hairs
- Trochanteral floccus
- Propodeal corbicula
- Females have a subgenal coronet
- Male with distinct, large gonobase

or hairs. Some are metallic blue or green, although usually darkly so, and still others are dark with gray hair bands and dark wings (note that this is not an exhaustive list of the color forms found in the genus).

Andrena are found worldwide except for Australasia and South America, but they are less common in tropical areas. Females are readily identified by the broad bands of short velvety hairs along the inner margins of the compound eyes, which reflect light quite strongly from some angles. Most *Andrena* species are solitary, although some are communal—sometimes with hundreds of individuals sharing a nest entrance. Most *Andrena*

species are active in the spring, with smaller numbers of species flying in summer or fall, and with a minority of species having two generations a year in some places. Overwintering underground as adults, some species dig up toward the soil surface while the ground is still very cold and may forage from forest-floor flowers in dense woodland before leafburst.

Andrena includes both specialist and generalist species. Some are economically important pollinators, especially of fruit trees, although their importance is often unacknowledged, with their fruit-producing activities often attributed to honey bees.

MEGANDRENA

BELOW | Male bees often sleep in flowers that their female counterparts might not visit during the day. In this case, a *Megandrena enceliae* is resting in a Ghost Flower (*Mohavea confertiflora*).

There are two species in this genus of southwestern USA bees. One, *Megandrena mentzeliae*, specializes on flowers of Spinyhair Blazingstar (*Mentzelia tricuspis*). While other specialists on this plant and its relatives might be found in most years, this one is very particular about the amount of rainfall it requires before it emerges and initiates nests. As a result, the bees often remain quiescent underground for more than a year and the number of inactive years is likely increasing as droughts become longer in duration due to climate change. The other species, *M. enceliae*, is less narrow in its tastes but seems to favor Creosote Bush (*Larrea tridentata*). Males may spend the night in large flowers that are not visited by the females for pollen during the day. The easiest way to tell the two apart (other than their floral preferences) is that *M. mentzeliae* has red markings on the metasoma, whereas *M. enceliae* does not.

GENUS
Megandrena

DISTRIBUTION
California to Arizona

HABITAT
Arid areas where the host plants are abundant

CHARACTERISTICS
- Facial fovea of female with velvety hairs
- Small, parallel-sided stigma
- Males have extensive pale markings on the face
- Gonobase a tiny ring to absent
- White apical hair bands on metasomal terga (sometimes medially interrupted in the red-marked *M. mentzeliae*)

EUHERBSTIA

There is only one species in this genus, the appropriately named *Euherbstia excellens*, which is a beautiful large bee with an orange-marked metasoma on a metallic blue background and smoky-and-orange wings. It is one of the most instantly recognizable bees in the world: no other bee looks like it. The males have cream-colored marks on the face; these are absent in the females.

The species is found in Chile, from the southern edges of the Atacama Desert to around Santiago. Nests have been found in cracked ground and on slopes with friable soil. It is abundant in December, in crucifer flowers along the road to the ski resorts southeast of Santiago, but as these plants are introduced, they cannot be the natural floral host for the species. It flies earlier in the austral spring the further north one goes in Chile.

LEFT | A male *Euherbstia excellens* obtains nectar from a Chile Nettle (*Loasa tricolor*), a plant that is very popular with local bees but inflicts pain on the unwary melittologist.

GENUS
Euherbstia

DISTRIBUTION
Central Chile

HABITAT
Semiarid habitats; recent records are from 2,600–6,200 ft. (800–1,900 m) in altitude

CHARACTERISTICS
- Blue-and-orange integument
- Female facial fovea weakly demarcated and lacking velvety hairs
- Claws of female with tiny inner tooth
- Large size
- Three submarginal cells
- Mandible of male lacking subapical tooth

ORPHANA

There are two species in this endemic Chilean genus, at least one of which has a melanistic form in which all the body hairs are blackish instead of shades of pale brown as in the image here. Females seem to fly mostly soon before sunset, while the ever-optimistic males start searching for mates much earlier in the day. Indeed, males fly so fast and for so long that by the end of the day they may be so exhausted that strong winds can simply blow them along the ground. At least one of the two species seems to specialize on Chile nettles (*Loasa* spp.). Nests of this genus have never been found, despite considerable effort, which involves slowly walking around near where the floral hosts and bees are found, and staring at the ground hoping to see a female fly back into its nest.

LEFT | *Orphana* bees also like Chile Nettles (*Loasa tricolor*), and the females of at least one species seem to specialize in collecting pollen from this plant.

GENUS
Orphana

DISTRIBUTION
Central Chile, from a few hundred miles north to a few hundred miles south of Santiago

HABITAT
Areas with soft soil where the host plant(s) are found, from coastal woodland to semiarid areas

CHARACTERISTICS
- Female lacking facial fovea
- Rather robust bees
- Three submarginal cells
- Female claws with large inner tooth
- Metasoma of males somewhat triangular

PROTOXAEA

There are three species in this genus, which comprises large bees with a black integument and, like other members of the subfamily, rather long, narrow wings. They are found in Mexico and southwestern deserts of the USA. The most commonly observed species, *Protoxaea gloriosa*, has extensive russet-colored pubescence and is frequently found on the yellow-orange flowers of *Kallstroemia*. The pollen of this plant is bright orange and the bees are often so completely covered in it that it is difficult to see the underlying body. However, it remains uncertain to what extent the bees visit this plant for pollen as opposed to nectar.

As with other genera in the subfamily, *Protoxaea* form vertical brood cells. The provisions supplied are runny and the larva swims on the surface of its food; it has a morphology that does not resemble that of larvae of other andrenids. Most of what we know about bee larvae has been discovered by American entomologist Jerry Rozen.

BELOW | A male *Protoxaea gloriosa* refueling before restarting his attempt to find a mate.

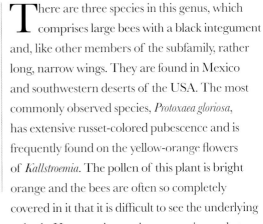

GENUS
Protoxaea

DISTRIBUTION
Southwestern USA and Mexico

HABITAT
Mostly semiarid areas

CHARACTERISTICS
- Stigma absent
- Marginal cell very long and narrow
- Labrum as long as wide
- Maxillary palpus present
- Male gonostylus present
- Male gonobase longer than wide
- Male S8 not emarginate apically
- Male T6 and female T5 lacking white lateral hair tufts

OXAEA

This is a genus of 10 species of large, very fast-flying bees. The males usually have brilliant metallic greenish bands on the apical impressed areas of the metasomal terga, while in females the entire metasoma is usually largely metallic.

As with many bees that fly rapidly and with males that search for females visually, the male's compound eyes are enormous. The ocelli are positioned low on the face to make room for the enlarged compound eyes, which are convergent above, and the first flagellomere is very long—as long as the scape—another feature commonly found in the fastest flying bees.

Nests are excavated in the ground and can be more than 6 feet (2 meters) deep. The brood cells are oriented vertically and the provisions left for the larvae are quite liquid, pooling at the bottom of the cell.

ABOVE | Here we can see a male *Oxaea* attending the base of the flower rather than going through the opening of the corolla. His mouthparts are probing the base of a flower, robbing nectar without performing any pollination.

GENUS
Oxaea

DISTRIBUTION
Mexico to southern Brazil and northern Argentina

HABITAT
Various, ranging from semiarid areas to moist forests

CHARACTERISTICS
- Stigma absent
- Mandible simple
- Marginal cell very long and narrow
- Labrum as long as wide
- Maxillary palpus absent
- Male gonostylus seemingly absent
- Metasoma with metallic markings
- Male lower face, antennal scape, and mandible usually with pale markings

RIGHT | This female *Oxaea flavescens* is taking a rest before continuing her search for a suitable nest site.

NOLANOMELISSA

This tribe and genus contain just a single species, but one with a remarkable evolutionary history. It diverged from its closest relatives before the dinosaurs (other than birds) became extinct more than 55 million years ago. Its sister group contains more than 1,400 species, so why, given that both lineages have had the same time to diversify, does *Nolanomelissa* include only one species? As its name suggests, this bee collects pollen and nectar from Chilean bell flowers (*Nolana* spp.). But here's another mystery: *Nolana* is a young genus, perhaps only 5 million years old, so the ancestors of this bee must have fed on something else for at least 60 million years. This species is known from a few hundred females but only three males—a third mystery is that we do not know why it has such a biased sex ratio.

BELOW | This female *Nolanomelissa toroi* has a full pollen load on her scopa, all of it obtained from the *Nolana patula* flower on which she was photographed.

GENUS
Nolanomelissa

DISTRIBUTION
A small area of the southern Atacama Desert around the Chilean city of Vallenar

HABITAT
Arid areas with healthy populations of *Nolana*, especially *N. rostrata*; not known from coastal areas or from altitudes above 8,200 ft. (2,500 m)

CHARACTERISTICS
- Three submarginal cells
- Long glossa and labial palpi with second palpomere longest and apical (fourth) palpomere minute
- Inner eye margins of female divergent below
- Lacking pale integumental markings
- Female T5 with apicomedian slit

ARHYSOSAGE

Most bees with a very pale integument are either very small or crepuscular, but neither is true of this small genus of eight species. *Arhysosage* are mostly yellow to pale brown in color with some black markings, although males are generally entirely yellow. The species are largely differentiated by details of their color patterns. The sizable, bent mandibles of the large-headed males are impressive and are used to hold on to females during mating, as the copulating pair fly from flower to flower conjoined. All species seem to be oligolectic on cactus flowers, and mating is usually initiated within a flower, but the male remains attached to the female as she forages.

BELOW | Despite their overall pale coloration, *Arhysosage* females are not usually well camouflaged on the cactus flowers they spend so much time visiting.

GENUS
Arhysosage

DISTRIBUTION
Southeastern Bolivia, Paraguay, southern Brazil, and northern Argentina

HABITAT
Semiarid and temperate Chaco and Pampas with cactus host plants

CHARACTERISTICS
- Female S6 relatively glabrous except for a marginal band of dense hairs
- Anterior tentorial pit in outer subantennal sulcus, at least in male, near midlength of sulcus
- Eyes divergent below
- Male mandible long and bent with preapical tooth
- Male penis valves complex
- Hind tibial spurs strongly curved apically
- Metasoma wider than thorax

CALLIOPSIS

BELOW | *Calliopsis* males often await mating opportunities on the flowers of their female's preferred plants—as this male *Calliopsis subalpina* is doing on a Desert Globemallow (*Sphaeralcea ambigua*) flower.

This genus of approximately 80 species used to have an amphitropical western hemisphere distribution, being widespread in North America and southern South America, but missing from southern Panama to northern Peru and most of Brazil. However, research published in 2021 suggests that the South American species should be classified separately. Thus, *Calliopsis* is likely only a North and Central American genus. Even as more narrowly defined, these bees are fairly diverse, ranging from small, dark, relatively unhairy bees to medium-sized species with quite long hairs. There are always pale markings on the face and sometimes on the metasoma. While aggregations of nests are known, the species seem to be solitary. As with other genera in its tribe, the males have complex elaborations to the structure of the penis valves, a somewhat uncommon feature among the bees.

One species from southwest US is unique in that males have falcate (hooked) forewings.

GENUS
Calliopsis

DISTRIBUTION
North America as far north as southern Canada, and Central America as far south as Panama

HABITAT
Semiarid areas or damper habitats with sandy soils

CHARACTERISTICS
- Female S6 relatively glabrous except for a curved, marginal band of dense hairs
- Anterior tentorial pit in outer subantennal sulcus, at least in male below the middle of the sulcus

- Male penis valve complex and elaborate
- Male metabasitarsus with hairs on dorsal margin at most half as long as metabasitarsus
- Female metatibia with keirotrichiate area much reduced

ABOVE | A mating pair of *Calliopsis puellae* standing on their floral host *Malacothrix*. Remaining *in copula* for a long time is a form of mate guarding, and towards the end of the day the male will stay connected with the female as they travel all the way back to the nest.

A genus of more than 70 species, these handsome panurgines often have spectacular yellow, orange, and/or red markings. Females of some species share nest burrows in communal societies.

Males often have remarkably enlarged heads, with large mandibles and sometimes with genal processes or horns on the face. These modifications might be associated with fighting over mating rights of males with females in the communal nests. Allometric variation, in which some body parts are scaled disproportionately with body size, seems to be unusually common in males of bee species in which the females nest communally. When the larger males have relatively enormous heads and associated weapon-like structures (mandibles, horns, etc.), this would suggest greater success in male-on-male combat.

GENUS
Psaenythia

DISTRIBUTION
South America, south of the Amazon River and east of the Andes except for one species in Chile

HABITAT
From semiarid areas to woodland

CHARACTERISTICS
- Anterior tentorial pit at junction of epistomal and outer subantennal sulci
- Episternal groove extending below scrobal groove
- Hind femur with longitudinal ridge above a glabrous longitudinal depression
- Meso- and metatibial spurs coarsely serrate
- Male S6 with apical concavity almost attaining base of sternum
- Male S8 lacking lateral apodemes

CAMPTOPOEUM

Species of *Camptopoeum* vary from being almost entirely black to extensively yellow, sometimes with a red metasoma. This genus forms one of an increasing number of Panurginae (some recently described) that have color patterns more commonly associated with wasps. *Camptopoeum* was moved from the tribe Panurgini to the Melitturgini in 2021. There are 31 described species, but in 2022 one will be separated out into a new genus that belongs back in the Panurgini, from which the other species in the genus have just been removed. These changes are all part of the ever-changing classification that hopefully more accurately reflects our increased understanding of bee evolution.

BELOW | The wings of a bee get somewhat frayed as it ages. This male *Camptopoeum frontale* has wings that are quite badly damaged, likely a result of a week or more of flying around looking for mating opportunities.

GENUS
Camptopoeum

DISTRIBUTION
Circum-Mediterranean, southern Europe, east to the Arabian Peninsula and Pakistan

HABITAT
Mostly semiarid regions within its geographic range

CHARACTERISTICS
- Two submarginal cells
- Episternal groove punctate, extending little below scrobal groove
- Male disk of S7 small with large apodemes
- Usually with extensive yellow markings

MELITTURGA

There are 17 species in this genus of relatively large, hairy panurgines. The male eyes converge above, a feature often seen in males that fly very fast as they search visually for similarly fast-flying females. As a result of this, the ocelli are lower down on the face than in most bees as there is no room for them in their usual position. The very long F1, longer than the scape, is another feature common to fast-flying bees, and the antennae of males are short overall, further supporting the idea that sight, rather than smell, is the main means of mate location. The short antennae make the males superficially resemble females, but their larger eyes and lack of a scopa enable separation of the sexes.

GENUS
Melitturga

DISTRIBUTION
Disjunct, being found from Portugal and Morocco to the Far East, and in southern Africa

HABITAT
Mostly semiarid regions within its geographic range

CHARACTERISTICS
- Stigma slender, somewhat parallel-sided, shorter than prestigma
- Three submarginal cells of approximately equal length along posterior margins

- Male eyes converge above
- F1 long in male (longer than scape)
- Male metatibia with a toothed dorsal carina
- Male lacking facial fovea

ABOVE | A male *Melitturga* in takeoff mode. The four apical foretarsomeres are somewhat expanded here, likely a sexually selected trait. This male has paler blue eyes than in the female.

MACROTERA

Bees in this genus used to be lumped together with the much more speciose genus *Perdita* (page 76), but are now separated based on various morphological differences and DNA data. In addition, *Macrotera* species retain the waxy brood-cell lining that is lost in *Perdita*. There are 31 species, found from western North America into Mexico.

One species has been studied in detail by American entomologist Bryan Danforth of Cornell University: *Macrotera portalis* females share their shallow nests with up to a couple of dozen other females and the nests can be reused over multiple years. Perhaps the most interesting aspect of the biology of this species is that males come in two discrete sizes—normal small males that fly around looking for mates on flowers, and large flightless macrocephalic males that are built like tanks, stay in the nest, and fight (sometimes to the death) with other big-headed males. The combats are over mating rights with females emerging from the same nest. Even if the female has already mated with a small male while foraging, the macrocephalic male will mate with her just before she lays an egg— perhaps making it more likely she will use his sperm. In some other *Macrotera* species the males have continuous variation in body size.

ABOVE | A female *Macrotera* on a *Sida* flower. Most Perditini are oligolectic, which means they forage for pollen on only one or a few closely related species of plant; indeed, some keys to their identification have couplets saying nothing other than the name of the plant the bees might be found upon!

RIGHT | A male *Macrotera latior* awaits the arrival of a potential mate to the flowers that she prefers.

GENUS
Macrotera

DISTRIBUTION
Western North America to southern Mexico

HABITAT
Semiarid habitats where the host plants are found

CHARACTERISTICS
- Anterior tentorial pit at junction of epistomal and outer subantennal sulci
- Marginal cell very short, extending little beyond apical submarginal cell
- Mandible with a narrow groove that curves upward toward base
- Episternal groove absent or short, not curving to scrobe

PERDITA

With almost 640 species, most restricted to southwestern USA and adjacent parts of Mexico, this genus forms a species swarm of mostly very small ground-nesting bees. One of its species, the aptly named *Perdita minima*, shares the record for being the smallest bee in the world at just ¹⁄₁₆ inches (1.6 millimeters) in length. Unsurprisingly, it favors similarly tiny flowers, such as those of Whitemargin Sandmat (*Chamaesyce albomarginata*). Many species are communal, with multiple females sharing a nest entrance. Unusually, the female covers the pollen ball with waterproofing rather than lining the brood cells. These bees are often so choosy in their host plants that identification keys sometimes use the host plant instead of morphological features to separate them (which isn't very useful if the specimen is caught in a trap).

GENUS
Perdita

DISTRIBUTION
North and Central America, from Alaska to Colombia, but with most species in southwestern USA and northern Mexico

HABITAT
Mostly in semiarid areas, but species from the north and east of the range occur in damper areas, including some that are submerged for part of the year (the fully grown larvae can "swim")

ABOVE | Many of the hundreds of species of *Perdita* look very similar to this one, and they are very difficult to identify to the species level.

OPPOSITE | *Perdita minima* shares the record for the smallest bee species. Here a female feeds on the tiny flowers of its *Euphorbia* host.

PANURGUS

There are three subgenera in the genus *Panurgus* (considered valid genera by some researchers). *Flavipanurgus* comprises species with yellow markings, as the name suggests (*flavo* being Latin for "yellow"). To date they have been found only in Spain and Portugal and include seven described species. *Simpanurgus* has been collected only once (in Spain) and no females have ever been seen. *Panurgus* comprises 35 entirely black species that occur throughout Europe to northern Africa, the Arabian Peninsula, and southern Russia.

RIGHT | A male *Panurgus banksianus* on Field Scabious (*Knautia arvensis*).

BELOW | Some males of *Panurgus* species have enlarged heads, even larger than in this image. Large-headed males are relatively commonly found in species that have communal nests.

GENUS
Panurgus

DISTRIBUTION
Europe including the Canary Islands, the Arabian Peninsula, and northern Africa

HABITAT
Semiarid areas to grasslands and field margins

CHARACTERISTICS
- Anterior tentorial pit at junction of epistomal and outer subantennal sulci
- Stigma wide
- Metapostnotum subequal in length to metanotum

- Male S7 without a midapical process
- Male S6 apically transverse or concave
- Episternal groove absent to short, at most attaining scrobal groove

An unusual feature of females of the subgenus *Panurgus* is that their scopal hairs are spiral or zig-zagged, which gives them a rather attractive appearance when looked at with sufficient magnification.

These are ground-nesting bees. At least some species are communal, with up to 20 females sharing a nest. Males often have enlarged heads.

HALICTIDAE

Halictidae is the second-largest family after Apidae, with well over 4,000 named species and hundreds awaiting description. The key diagnostic feature that helps determine a bee as a halictid is obscure: instead of being a hairy flap near the base of the stipes (as in the figure on page 21), the lacinia is a finger-like, largely hairless lobe that is situated further up the maxillary tube. More easily identifiable characteristics for distinguishing many species include a narrow pseudopygidial area on T5 of females and a basal vein that is strongly curved, especially toward the base.

Halictidae has a worldwide distribution and includes four subfamilies. The Rophitinae can be recognized by having antennae low on the face. In a few species the structure of the labial palpus resembles that of a long-tongued bee and a few species seem to have two subantennal sulci. The 250-plus species are divided among 10–12 genera, all but two with a restricted geographic distribution. These ground-nesting bees are absent from Australasia, and most species have very narrow floral preferences.

The Nomiinae can be readily separated from other halictids by the combination of a wider space between the antennal socket and clypeus and by the absence of the episternal groove beneath the scrobe. In those species with three submarginal cells (the majority), the second cell is the smallest and the third the largest. There are more than 600 species in this subfamily, found on all continents except Antarctica and South America.

The Nomioidinae is a small subfamily with three genera and fewer than 100 described species of small bees with extensive pale markings, often on a metallic background. Their ground nests are sometimes occupied by multiple females in a seemingly communal society. They are found from southern Europe to Australia, including Africa. The combination of antennal sockets distant from the clypeus, a distinct episternal groove, and the lack of a pseudopygidial area suffices to identify these bees among other subfamilies of Halictidae.

The Halictinae is the largest halictid subfamily, with around 50 genera and more than 3,500 species. It is global in distribution except for Antarctica. Most species nest in the ground, but some of those that nest in soft wood or stems are crepuscular or nocturnal. They have a remarkable diversity of social organization: many are solitary, others are communal, and many are eusocial, with colonies ranging in size from two bees (a queen and one worker) to perhaps 1,000 individuals. One species has queens that live for five years (twice as long as an average honey bee queen), while other eusocial species have colonies that last just one season (although nests may be reused). Some species have semisocial colonies that are not just a stage during the development of a eusocial society. Most species have a narrow pseudopygidial area on the T5 of females (the exceptions are cuckoo and socially parasitic species, which usually have at least a hint of this area) and in both sexes a basal vein that is strongly curved, especially toward the base.

This genus contains 170 species and occurs throughout the northern hemisphere. Most species are oligolectic—collecting pollen from one or a few closely related plant species. The common northern North American species *Dufourea novaeangliae* collects pollen only from Pickerelweed (*Pontederia cordata*), an aquatic plant, and thus is one of the few bees best observed from a canoe.

Most species are rather drab dark brown to black, but some are a dull metallic green or blue—rarely quite brightly so—and the most attractive ones combine a metallic head and mesosoma with a red metasoma.

Males possess a wide range of secondary sexual characteristics, especially modifications of the hind leg, and sometimes the midleg and/or antenna. They also have patches of hair forming an extensive beard and elsewhere on the body, giving them an unevenly fluffy appearance.

RIGHT | *Dufourea maura* is a very dark bee; even the wings are dark.

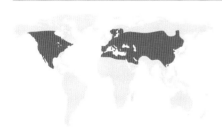

GENUS
Dufourea

DISTRIBUTION
Throughout the northern hemisphere except the coldest areas

HABITAT
From arid areas to damp meadows at a wide range of altitudes, and with different species occupying different ecological zones depending on their floral hosts

CHARACTERISTICS
- Antennae low on the face
- Clypeus protuberant
- Usually with two submarginal cells
- Male without a pygidial plate, although sometimes with a shiny bare area
- Male with base of S8 strongly concave, spiculum absent
- Labial palpus with second palpomere short like the third and fourth

BELOW | It is not surprising that, as bindweed specialists, both the female *Systropha* here and the male shown opposite were photographed on a hostplant flower.

The spiral-horned bees are unique in that the apex of the male antennae is made up of flagellomeres that are reduced in size and curled, sometimes forming a hook. Most male bees have 11 flagellomeres, but in spiral-horned bees this number is as low as eight in some species. The males are also adorned with a wide range of other secondary sexual characteristics, ranging from crests to almost hatchet-shaped extensions on the metasomal sterna, lateral spines on the terga, and swellings on the ventral surface of the thorax. Females are also somewhat unusual in that they collect pollen over a larger proportion of their body surface than do most bees, with not only the hind legs and the metasomal sterna involved, but also much of the metasomal terga. The females collect pollen exclusively from species of Convolvulaceae. Both sexes have heads that look rather small for the size of the rest of the body.

GENUS
Systropha

DISTRIBUTION
From Spain north to the Baltic states, east to Sri Lanka and Thailand, the Arabian Peninsula, and parts of Africa

HABITAT
Semiarid areas around the Mediterranean and elsewhere to tropical Asian forests, wherever good populations of their floral hosts, Convolvulaceae, can be found

CHARACTERISTICS
- Antennae low on the face
- Males have apical flagellomeres at least somewhat reduced and curled
- Females have pollen-collecting hairs on the metasomal sterna and terga as well as on the hind leg
- Head rather small for the size of the body

The genus contains 30 described species and can be found over much of the eastern hemisphere except Australasia. Its occurrence in Pakistan, the extreme north of India, and Sri Lanka suggests that there remain undiscovered populations in India. The bees nest in the ground and their brood cells are at the end of lateral branches that are barely any longer than the length of the brood cells.

BELOW | The spiral nature of the male antenna is clearly seen in this image of *Systropha curvicornis* and can be contrasted to the straight antenna of the female that is shown opposite.

XERALICTUS

BELOW | A male *Xeralictus* sits on a flower of Desert Rock Nettle (*Eucnide urens*), perhaps in wait for a female to visit for floral resources.

There are only two species in this genus of unusually large bees for the subfamily. Both are spring species that are found in the deserts of southwestern USA and adjacent parts of Mexico. The females collect pollen only from flowers of the family Loasaceae, with one of the two species concentrating on the cream-colored silky flowers of Desert Rock Nettle (*Eucnide urens*). This plant has long, dense stamens, and the females of *Xeralictus bicuspidariae* bury their front ends among them such that only the metasoma is visible. The females of this species are dimorphic in color, with a metasoma that is either dark brown or red. Females of a small *Perdita* species (page 76) have been observed trying to steal pollen from the scopae of the larger *Xeralictus* females.

GENUS
Xeralictus

DISTRIBUTION
Southern California, Nevada, and Arizona, USA; and Baja California, Mexico

HABITAT
Hot deserts within its range wherever the host plants grow in abundance

CHARACTERISTICS
- Large halictines with antennae low on the face
- Labial palpus no longer than prementum
- The midtibial and anterior hind tibial spurs are coarsely serrate
- Male mandibles and S4–S6 are strongly modified
- Females have an apically concave pseudopygidial area

DIEUNOMIA

Bees in this genus come in two forms: large, very black bees; and medium-sized bees with red markings and at least partial white hair bands. Each has its own subgenus, with members of the nominate subgenus being the darker form and those of the subgenus *Epinomia* the red-marked form. The two subgenera have five and four species, respectively. Females are unusual among bees in that the scopa extends onto the sides of the metasomal terga.

These bees specialize on pollen from Asteraceae. They may survive even in agricultural fields of sunflowers, as their vertical brood cells, where the fully grown larvae overwinter, are placed deeper than the farm machinery plows.

RIGHT | Pollen collects over a larger proportion of the body surface of *Dieunomia heteropoda* bees than is usual, as can be seen in this image.

GENUS
Dieunomia

DISTRIBUTION
From southern Canada to parts of Central America

HABITAT
Found especially where there is sandy soil with abundant Asteraceae nearby

CHARACTERISTICS
- Three submarginal cells, the second one considerably smaller than the others
- Midtibial spurs unmodified
- Marginal zones of T2–T4 with hair bands
- Female with scopa on metasomal sterma as well as hind leg, extending onto sides of terga
- T1 with a medial longitudinal depression on the anterior sloping surface, this margined by a raised area
- Gena and vertex well developed, with the head thicker than in relatives

G rasses are wind-pollinated plants with tiny, protein-poor pollen grains. It is therefore surprising to find any bees that collect grass pollen, but some species in the genus *Lipotriches* do just this. Pollens of most plants are covered in a sticky coating that makes it easy for them to adhere to the body of a pollinator. In grass pollen this coating is much reduced. As a result, bees that collect grass pollen have to forage when the air is humid, and even in the wet tropics this means foraging early in the morning. *Lipotriches* bees are thus mostly active very early in the day.

LEFT | A female *Lipotriches hylaeoides* collecting pollen from the grass *Sporobolus pyramidalis*.

RIGHT | A male *Lipotriches australica*. This species and that shown to the left are considered to belong to different genera by some researchers.

GENUS
Lipotriches

DISTRIBUTION
Most of Africa, the Arabian Peninsula, and east and south through SE Asia and Australasia

HABITAT
Diverse, from semiarid to warm, moist tropical areas

CHARACTERISTICS
- Lacking pale metasomal bands, although sometimes the metasoma is mostly reddish
- Three submarginal cells
- Tegula usually not considerably expanded
- Pronotum with a complete or almost complete transverse carina

As with other nomiine genera, *Lipotriches* has undergone substantial taxonomic change and its classification is still not universally agreed upon, with the result that it is uncertain how many species belong to the genus, although the number ranges from 123 to more than 340. These bees are black to brown, sometimes with an orange or red metasoma. Males of some species, especially those with a narrow first metasomal segment, have unusual patterns of knob-like bristles on S5, the patterns of which are species specific. Most of the other sterna also have secondary sexual characteristics in males, with ridges, spines, or modified hairs.

BELOW AND OPPOSITE | A female *Nomia melanderi* on Alfalfa (*Mendicago sativa*), the floral host for which it is a managed pollinator.

Bees in the genus *Nomia* are called pearly-banded bees because most species have orange, yellow, white, or opalescent bands on the apical impressed areas of three or more of their metasomal terga. Only one other bee genus has similar opalescent bands, and its name—*Nomiocolletes*—tells us that it is both somewhat *Nomia*-like and that it is from a different family, the Colletidae. There is no chance of confusing the two, though, because the latter is found only in South America and *Nomia* is found almost everywhere except South America.

There is disagreement among experts as to how to classify these bees, with some including a wider range of bees in the genus than do others. At most, there are around 130 species. One of these, the Alkali Bee (*Nomia melanderi*), is the only ground-nesting bee that is managed at a large scale. It is used in parts of northwestern USA for alfalfa pollination.

GENUS
Nomia

DISTRIBUTION
Worldwide except south of Mexico, cold temperate areas, and some small islands; introduced to New Zealand

HABITAT
Diverse, from semiarid areas to tropical moist areas

CHARACTERISTICS
- Three submarginal cells, the second one considerably smaller than the others
- White, yellow, orange, or opalescent bands on the metasomal terga
- Marginal zones of T2–T4 lacking hair bands

Its common name derives from the nest-site choice of the bee—alkaline salt flats. New populations can be seeded by taking a block of soil containing nests to a new location. In areas where the species is used, road signs may suggest a slow driving speed to prevent too many bees being killed on windscreens.

PSEUDAPIS

S ome bees have taken sexually selected traits to absurd
levels, with females selecting for extreme modifications
of many body parts of their suitors. *Pseudapis* is one genus that
exhibits the results of such selection on numerous body parts,
each of which comes in a remarkable array of forms: most
males have massively swollen hind femora, scythe-like
extensions of the hind tibiae, and huge teardrop-shaped hairs
that look as if they are made of plastic on the ventral surface
of the hind legs and sometimes on adjacent parts of the body.
As in most bees, the male genitalia are strongly divergent
among related species, and as in most Nomiinae, the
metasomal sterna bear ridges, spines, or odd clusters of hairs.
Particularly interesting are the black semaphore flag–like
modifications of the apical tarsomere of the midleg in some
males. These come in two forms—either as fringes of thick
hairs on both margins, or the tarsomere itself becomes
disk-shaped. It seems that the males have evolved two solutions
to the same female preference.

Another unusual feature of these bees, found in both sexes,
is the remarkably expanded tegula. This is something that seems
to have arisen multiple times among nomiines but almost never
among other bees. With 50 species found throughout much of
the eastern hemisphere, this genus includes most bees that
display giant tegulae.

GENUS
Pseudapis

DISTRIBUTION
Throughout Africa and from southern
Europe to west of Wallace's Line, but
absent from Madagascar

HABITAT
Diverse, from semiarid areas
to the wet tropics

CHARACTERISTICS

- Nomiinae with considerably expanded tegulae
- Pronotal lobe with a large lamella
- Mesoscutum anteromedially strongly downcurved, often carinate
- Preoccipital carina present

ABOVE | The large kneecap-like tegula is clearly seen at the base of the wing of this male *Pseudapis diversipes*.

Pretty little metallic and yellow bees, *Nomioides* seem to be the ecological equivalent of the North American *Perdita* (page 76) or the South American *Xeromelissa* (page 137), in that they are small yellow-marked bees that can be quite common under the right conditions. However, the extent to which they are as oligolectic (as these other genera often are) remains uncertain.

These tiny bees have their own special, perhaps even tinier, kleptoparasites, the bee genus *Chiasmognathus* (Apidae: Nomadinae), which was described in 2006. *Nomioides* nest in the ground, sometimes occupying the vacated brood cells of larger bees. Multiple females can be found within the same nest, but it seems they lack castes and are likely communal. There are 55 species.

LEFT | The dark metallic orange and pale yellow patterns on these *Nomioides* females are attractively set off against the blue petals and pink pollen of a host plant, Sheep's-bit (*Jasione montana*).

GENUS
Nomioides

DISTRIBUTION
From Portugal to Sulawesi, including much of Africa and many islands

HABITAT
Semiarid to arid areas with sandy soil

CHARACTERISTICS
- Episternal groove distinct and extending beyond scrobal groove
- Female labrum with apical process that lacks a keel
- Female prepygidial fimbria not divided medially
- Marginal zone of T2 translucent
- Anterior tentorial pit at the end of an epistomal lobe
- Inner eye margins angularly concave
- S8 of male with apical process much longer than broad

AUGOCHLORA

With almost 120 species, this is a diverse genus of usually brightly metallic species. The common North American species imaged here, *Augochlora pura*, is bright green to bluish, although populations in Florida are bright purple. Species elsewhere range widely in color, from red through to purple, although some are mostly black.

Augochlora pura is a solitary species that nests in wood, especially rather soft wood that can be manipulated in a similar manner to soil. Thus, its brood cells have a consistency resembling sawdust. Research in 2010 demonstrated that the makeup of the brain of this species differs from that of a social relative.

LEFT | The spectacular metallic colors found on most bees of the tribe Augochlorini are clearly on display on this foraging female *Augochlora pura*.

GENUS
Augochlora

DISTRIBUTION
North, Central, and South America and the Caribbean except colder regions of the far north and far south, and west of the Andes south of Peru

HABITAT
Diverse, from deserts to rain forest

CHARACTERISTICS
- Bright metallic sweat bees
- Pseudopygidial area of female with a slit
- T7 of male not recurved at apex
- Prementum not unusually narrow
- Female with inner hind tibial spur teeth short and rounded
- Epistomal lobe acute
- Marginal cell truncate

AUGOCHLORELLA

A taxonomic revision is the most useful work a taxonomist can publish, because it gathers data from as many specimens as possible, compares identifications for all of them, and summarizes all available information. This involves a considerable amount of work and, as a result, not many genera receive multiple revisions. *Augochlorella* is an exception, having been revised in the 1960s, again in the 2000s, and then studied in detail again in 2019. However, despite this repeated attention there remain some taxonomic complexities that need ironing out.

The genus contains 19 described species found throughout much of North, Central, and South

GENUS
Augochlorella

DISTRIBUTION
Throughout North, Central, and South America, but absent in the far north, west of the Andes, and in the Caribbean islands

HABITAT
Diverse, from arid to forested habitats, but not abundant in the Amazon Basin

CHARACTERISTICS
- Bright metallic
- Pseudopygidial area of female with a slit
- T7 of male not recurved at apex
- Prementum not unusually narrow

LEFT AND ABOVE | The name *Augochlorella aurata* could be translated as "little golden-green gilded," which seems an apt description of this species. Individuals vary from mostly green (as in the female on the left) to almost entirely coppery (as in the male on the right).

- Female with inner hind tibial spur teeth short and rounded
- Epistomal lobe obtuse or right-angular
- Marginal cell acute to narrowly truncate

America. It includes the one imaged here, *Augochlorella aurata* from North America. This species is normally social, with each nest in summer containing a queen and a few workers. Nests are underground, often with a turret at the entrance, and the brood cells are gathered together in a cluster surrounded by a cavity. At the northern edge of the species' range, the summer is too short for the predictable rearing of a worker brood followed by males and the next year's potential queens. Consequently, some females nest entirely solitarily, whereas others produce only one or two workers for an average of just 0.5 workers per nest-founding female.

MEGALOPTA

These tropical bees are well known for their nocturnal habits and ability to fly in very low light conditions. As with other bees active in dim light, they have enlarged ocelli that almost fill the space between the compound eyes toward the top of the head. Also, as with many other nocturnal bees, their body is often pale in colour, although the one imaged here has the pale colour restricted to the legs and metasoma.

There are about 30 species, all of which nest in twigs hollowed out by the female, which initiates the nest. They are socially polymorphic bees, with some females remaining alone and others producing a small brood of 1–3 workers. It seems that the queens manipulate their daughters into staying at home to help them produce more offspring.

RIGHT | The unusually large ocelli that characterize members of the genus *Megalopta* can be seen on the top of the head of this individual from Ecuador.

GENUS
Megalopta

DISTRIBUTION
From Mexico south to northern Argentina and southern Brazil

HABITAT
Tropical areas with relatively moist forests

CHARACTERISTICS
- Ocelli enormous
- T5 of female slit
- T7 of male not recurved at apex
- Greenish to bronze colors on the head and parts of the thorax; metasoma often pale brown
- Posterior hind tibial spur with few long, well-separated teeth

- Mandible with subapical tooth on both the dorsal and oral surface
- Hamuli in one dense cluster

AGAPOSTEMON

Male *Agapostemon* are among the most easily identified bees on the planet—the head and thorax are usually bright metallic green (sometimes blue, rarely black or orange) and the metasoma has yellow bands on a black or metallic (rarely orange or red) background. Females are either all metallic or are metallic with a black to orange metasoma, although a few can be entirely black. It seems that metallic versus black coloration is a genetic polymorphism in some species, with heterozygotes being very dark blue.

There are 41 described species of these ground-nesting bees. Some are solitary while others are communal, with females sharing a common

LEFT | Any North American bee that is green in front and has yellow and black bands at the back is instantly recognizable as a male *Agapostemon*.

RIGHT | And in eastern North America, any bee that is green at the front and black with white hair bands like those seen here is a female *Agapostemon virescens*.

GENUS
Agapostemon

DISTRIBUTION
Caribbean and North, Central, and South America, from southern Canada to northern Chile

HABITAT
Varied, from deserts to lush fields and woodland clearings; some species are abundant in cities

CHARACTERISTICS
- Pseudopygidial area narrow, lacking a slit
- Terga lacking apical hair bands
- Posterior surface of propodeum surrounded by a distinct carina
- Males with yellow bands to the metasoma

entrance to the main burrow but with their own lateral branches beneath. They construct their own brood cells, collect their own pollen, and lay their own eggs. The advantage to this arrangement seems to be that the frequent traffic of females coming in and out of the nest deters potential enemies, making access to the brood cells and the food stored within them more difficult. One communal species is known to let "strangers" into its nests; thus, safety in numbers seems to be the motto for the communal species, while the solitary ones have to take extra effort in making their nest entrances difficult to find. Some *Nomada* cuckoo bees (page 194) are much more successful at laying eggs in *Agapostemon* nests that have only one female than in those that are communal.

CAENOHALICTUS

GENUS
Caenohalictus

DISTRIBUTION
From central Mexico to Patagonia

HABITAT
Diverse, from arid to humid regions, and from low altitudes to the high Andes

CHARACTERISTICS
- Female pseudopygidial area entire
- Male T7 strongly recurved apically
- T2–T4 lacking basal hair bands and lateral graduli
- Eyes hairy
- Metapostnotum long, with granular sculpture

LEFT | *Caenohalictus* species come in a range of sizes and colors, although most are green like this foraging female.

BELOW | One of the characteristics of *Caenohalictus* is that their eyes are hairy, as can be seen in this female silhouetted against a pale pink flower petal.

There are 56 described (and many undescribed) species in this genus, but their overall appearance is very variable and it seems probable that it will eventually be divided into several genera. Most are metallic in color, ranging from orange through green to blue and purple, often with red highlights, but some are brown and orange, and others are dull black with long white hairs, making them appear gray. One feature they all share is unusually hairy eyes. In honey bees the hairs serve to detect the direction of the wind and assist with navigation in windy conditions. Why this feature shows up scattered among other bees, albeit uncommonly, is not known, but it does turn up in most families at least once and not necessarily in species inhabiting particularly windy places.

These bees nest in the ground solitarily or in groups of communal females that share little more than a nest entrance, presumably for similar reasons as for *Agapostemon* (page 100).

HALICTUS

BELOW | Some taxonomists divide *Halictus* into two genera, one of which contains metallic species like this female *Halictus* (*Seladonia*) *subauratus.*

The word *Halictus* is based upon the Greek word meaning "gregarious." This is a suitable name for these bees because they tend to nest in large aggregations. Each nest may contain a small society, usually in the form of a small eusocial colony, although many such species might have overwintered sisters that share a nest in spring, with one of them becoming queen and her siblings working for her.

GENUS
Halictus

DISTRIBUTION
Worldwide, but absent from Australasia, the islands of SE Asia, and much of South America; most numerous in the northern hemisphere and most diverse in southern temperate parts of Eurasia

HABITAT
Diverse, but less common in rain forests than in meadows and dry scrub

CHARACTERISTICS
- Distinct pseudopygidial area that is not slit medially
- Male T7 strongly recurved apically
- Three submarginal cells, with the veins surrounding them all strong
- Apical bands of pubescence on the metasomal terga; those with the entire terga covered in pale hairs have an unusually short metaspostnotum

Many species are black, some are greenish, and all have apical hair bands on the metasomal terga. However, in some the abdomen is covered in short, pale appressed hairs, making the bands hard to discern.

Most of the 200-plus species are from temperate regions ranging from Ireland to Japan, but *Halictus* also extends into North America, with a few green species reaching South America as far south as Brazil. Green species can also be found throughout Africa and parts of tropical Asia. The Brazilian species might have the largest caste dimorphism of any bee, with workers as small as half the size of the queen, such that it is possible a queen might be able to collect enough pollen for the development of a daughter from just one foraging trip.

ABOVE | *Halictus rubicundus* is one of the most widespread wild bees in the world, being found across North America and from Britain to Japan.

LASIOGLOSSUM

This is the largest genus of bees, with almost 1,900 described species. Most are small and predominantly black, brown, or dull metallic in color, but some have a red metasoma and a few Australian species are entirely pale brown and others more brightly metallic. A handful of North American species are nocturnal, flying when their floral hosts—such as evening primroses (*Oenothera* spp.)—are in bloom, but most species are seemingly generalists.

Lasioglossum contains the widest range of social behaviors of any insect genus. Many species are entirely solitary, while some are communal, with up to hundreds of individuals sharing a nest. Still others are eusocial, sometimes with queens and workers of nonoverlapping sizes (sometimes workers were originally described as a different species from the queens), and with colonies starting from a single foundress or several sisters in a semisocial society in which one female lays eggs and her sisters forage. One species, *Lasioglossum marginatum*, has queens that live for 5–6 years and produce one brood of workers each spring, starting with just 2–3 but eventually with

GENUS
Lasioglossum

DISTRIBUTION
Worldwide except for Greenland, Iceland, some Arctic islands, and Antarctica

HABITAT
Almost all habitats, including salt marshes and vernal desert pools

CHARACTERISTICS
- Distinct pseudopygidial area that is not slit medially
- Male T7 strongly recurved apically
- Usually three submarginal cells, with at least the apical veins weakened in females
- Bands of pubescence on the metasomal terga are apical in the nominate subgenus but may be basal in some others

hundreds. To make matters even more complicated, some species are cuckoo bees and others are social parasites, with females coexisting within the host colony. Truly, a whole book of this length could be written about just this one genus.

ABOVE | This female *Lasioglossum* (*Dialictus*) is intent on obtaining nectar rather than pollen from this flower.

OPPOSITE | With so many species in the genus *Lasioglossum*, it is not surprising these bees are common. They are easily identified to genus level, but with hundreds of species looking just like this one, obtaining a species-level identification usually requires an expert.

There are almost 300 described species in this genus of cuckoo bee. They mostly attack bees of the genus *Lasioglossum* (page 106), but some attack *Halictus* (page 104) or other Halictidae, while a few attack bees from other families. While some *Sphecodes* species have a range of hosts, individual bees concentrate on just one host species. Given that their main hosts are almost global in distribution, it is not surprising that this genus also occurs almost everywhere bees can be found.

Species that attack solitary hosts have a relatively easy task of entering a nest and ovipositing on a host pollen ball while the nest owner is out foraging. However, those that attack social hosts have a more complex task. One European species attacks the nests of a host species that are well defended with numerous workers, including a guard. The cuckoo bee attacks when the soil is moist and digs toward the main burrow from the side, thereby cutting off the guard. The robustly built cuckoo bee can then sting and bite at other workers on her way to the booty produced by the host's activities. At least some species have a very simple surface chemistry compared to their hosts, perhaps making it less likely they will be detected.

These bees are mostly black with a red metasoma, although some are all black and a few have red markings on the thorax.

ABOVE | *Sphecodes* females, such as this *S. albilabris* female, spend a lot of time on, or close to, the ground looking for host nests.

OPPOSITE | Most *Sphecodes* species are black with blood-red markings on the metasoma, the latter giving them their German common name of Blutbienen—"blood bees."

GENUS
Sphecodes

DISTRIBUTION
Worldwide except for Greenland, Iceland, some Arctic islands, Antarctica, and most of Australia

HABITAT
Diverse, occurring wherever there are good host populations

CHARACTERISTICS
- Three submarginal cells with all surrounding veins strong
- Head very wide
- Relatively glabrous and strongly sculptured
- Females lack a scopa
- Female mandible with subapical tooth

STENOTRITIDAE

This is the smallest bee family, with just two genera and 21 species. They are restricted to Australia, with most species found in the western half of the continent. They are large, fast-flying bees, and are the only Australian bees with two subantennal sulci. As with the Andrenidae, these define the subantennal sclerite, but unlike andrenids, the Stenotritidae subantennal sclerite is triangular (see page 2), because the two sulci meet before attaining the epistomal sulcus. Another identification feature is a rounded apex to the glossa. Stenotritidae nest in the ground, with nest depths varying from 6 inches (15 centimeters) to more than 10 feet (3 meters).

The two stenotrid genera are easily differentiated. In *Ctenocolletes*, the males have a pygidial plate that is defined by carinae, and a simple structure to S7 that has no distinct apical lobes and very short basal apodemes. In contrast, *Stenotritus* males lack a pygidial plate, and although the area where one would be is bare, there are no carinae surrounding it. In addition, they have an almost X-shaped S7 with long basal apodemes and apicolateral lobes. Females of the two genera are distinguished by the shape of their posterior hind tibial spur: in *Ctenocolletes*, this is broad for about the basal third of its length and then abruptly narrows, and it bears long, coarse teeth; and in *Stenotritus*, the spur gradually narrows from the base to the apex, and the teeth are finer and shorter.

CTENOCOLLETES

There are 10 species in the genus *Ctenocolletes*, most of which are black with hairs ranging from black to gray and orange, often with narrow, pale hair bands on the metasomal terga. Some species have a bright orange prepygidial fimbria, one has bright yellow integumental bands, and another has a largely orange metasoma. But with its metallic green body surface and attractive patterns of white and black hairs, *Ctenocolletes smaragdina* is arguably the most beautiful bee in all of Australia. These are large (⅝–¾ inches/15–20 millimeters long), remarkably fast-flying bees, and melittologists will often swing their nets in vain trying to catch them— although as mating pairs often remain coupled during flight, both sexes can sometimes be caught with a single swipe.

These bees nest in the ground, often in deep sand, making large tumuli at the nest entrance.

BELOW | This *Ctenocolletes nicholsoni* male is a beautiful bee even if it is not as gaudy as its relative on page 110.

GENUS
Ctenocolletes

DISTRIBUTION
Australia, primarily the western half

HABITAT
Arid and semiarid areas

CHARACTERISTICS
- Female labrum with undivided basal swelling
- Female posterior metatibial spur widest near midlength
- Male S7 a simple transverse band lacking lobes

STENOTRITUS

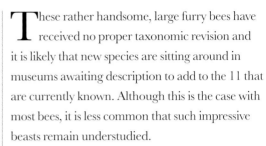

These rather handsome, large furry bees have received no proper taxonomic revision and it is likely that new species are sitting around in museums awaiting description to add to the 11 that are currently known. Although this is the case with most bees, it is less common that such impressive beasts remain understudied.

Like species of *Ctenocolletes* (see opposite), the other genus in the family, these are fast-flying bees. The males have remarkably large eyes, as is typical for bees that search for fast-flying females. In some species the males will hover, then move a little, and then hover some more. They will often travel the same route, repeatedly awaiting for females, but they remain almost impossibly fast to catch.

Stenotritus bees nest in harder soil than do their closer relatives, and their nests are also shallower.

BELOW | A male *Stenotritus* in flight.

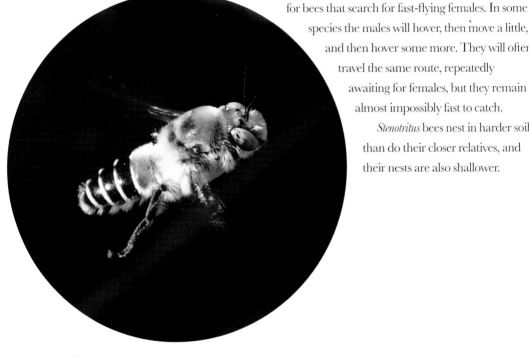

GENUS
Stenotritus

DISTRIBUTION
Throughout Australia except the far north and east

HABITAT
Arid and semiarid areas

CHARACTERISTICS
- Female labrum with divided basal swelling
- Female posterior metatibial spur widest near base
- Male S7 with long basal apodemes and long, hairy apical lobes

COLLETIDAE

This large and diverse family of bees is divided into nine subfamilies. In superficial appearance, they span the entire range of morphotypes found among all bees, with narrow wasp-like species, tiny fairy bees, normal honey bee–like bees, and large, fat furry species. What unites these diverse groups is the truncate, concave, or bifid glossal apex (males of a few Australasian species retain a pointed glossa) and the habit of lining their brood cells with a cellophane-like material.

Paracolletinae, as currently understood, is the smallest colletid subfamily, with just one genus and eight species. They include ground-nesting honey bee–sized Australian species. Australian entomologist Terry Houston recently discovered that the fully grown larvae spin a cocoon, which adds support to a relationship with the Diphaglossinae, the only other colletids that spin cocoons (although one colletine also does this). It has also been suggested that these bees be united within Diphaglossinae as a single subfamily.

Diphaglossinae comprises mostly large, very hairy bees with the stigma smaller than the prestigma and the glossa deeply bifid. At least some provide their offspring with a soupy food mass that ferments within the brood cell. The 130 or so species are found in South and Central America, with a few reaching the southern USA and the larger Caribbean islands.

Some fly in dim light conditions, often around sunrise and/or
sunset, and some are truly nocturnal.

Neopasiphaeinae is an interestingly diverse subfamily of almost
450 ground-nesting bees with a primarily Gondwanan
distribution. They can be identified as colletids with pygidial
plates and a prepygidial fimbria in females, the prementum
lacking a fovea, and, in all except a few instances, without a
deeply bifid glossa. Females usually have a basitibial plate,
which in males is often absent.

Callomelittinae is another small subfamily endemic to Australia,
along with one species known from New Caledonia. They are
coarsely sculptured, mostly red-and-black bees, and the mandible
has three even-sized teeth for excavating nests in wood.

Colletinae is a group of four genera and almost 550 species
restricted to South America, except for the most speciose genus,
which is found everywhere except Australasia and the polar
regions. These are relatively hairy bees that lack pygidial and
basitibial plates and have no prepygidial fimbria. Almost all nest
in the ground, except for a few southern South American species
that nest in hollow twigs or branches.

Scraptrinae also contains just one diverse genus, but with almost 70 species. They are almost entirely restricted to southern Africa. These bees nest in the ground and are the only African colletids with both a premental fovea and a scopa.

The Euryglossinae are mostly relatively bald and small, with a very short, broad, truncate glossa; many species have an entirely pale integument. Like the Hylaeinae, they carry pollen back to their nests inside their digestive system rather than on scopal hairs. Other than a couple of extralimital introductions, they are endemic to Australasia, where they can be abundant, especially when eucalypts are in flower. There are more than 400 described species, with nest sites ranging from preexisting cavities to excavations in the ground. This subfamily can be identified as colletids that lack a scopa but have a narrow pygidial plate and an unmargined spiculate area apically on the prementum.

The Xeromelissinae are small, not very hairy bees. Females are most easily recognized as colletids with a scopa that is at least as well developed on S2 as it is on the hind leg (one exception has almost no hairs on S2). It includes some species with extreme morphologies of sexually selected traits in males and others with unusual adaptations to floral hosts in females. Xeromelissinae are from Central and South America, along with one species in the Lesser Antilles. More than 150 species have been described, although there is perhaps a similar number awaiting description.

The Hylaeinae are the masked bees. Males of four of the genera are unique among Colletidae in having a pointed glossa. There are seven genera in total, all but one of which is restricted to Australasia; the exception is almost cosmopolitan. Some Australasian species are brightly metallic and some are unusually fast fliers compared to species elsewhere. Most nest in twigs and other hollow plant material, including plant galls, although some nest in the ground. Hylaeines can be identified as colletids that lack a scopa but have an extensive and distinctly margined premental fovea and no pygidial plate.

CAUPOLICANA

There are 38 species in the genus *Caupolicana*, all of which are attractive large, densely hairy bees. The genus ranges from Patagonia to central USA and also includes some Caribbean islands. Many species are active very early in the morning, and some even fly at night. They all nest in the ground, with wide, mostly shallow, burrows leading to large brood cells that are oriented vertically. The food the mother provides is (as is often the case in bees of this family) more liquid than for most, which is made possible by the waterproof, cellophane-like brood-cell lining. The sweet, soupy provisions ferment in the brood cell, the entrance to which is curved, something like a simple fermentation lock. Many species nest in soft sandy soil and the old cocoons can often be found as the nest site erodes.

GENUS
Caupolicana

DISTRIBUTION
South America, Central America, Caribbean islands, and southern USA as far north as Kansas, Colorado, and Utah

HABITAT
Widespread, from sea level to at least 13,000 ft. (4,000 m); many species prefer to nest in sandy soils

CHARACTERISTICS
- Deeply forked tongue
- F1 at least almost as long as scape
- Episternal groove complete
- Marginal cell lacking basal narrow extension
- Female second metatarsomere longer than wide
- Male posterior metatibial spur not fused to tibia

Males often aggregate in large numbers around nest sites where females might emerge, and may hover over a nest entrance waiting for a female to leave the nest. Such hover-and-dart mate-searching strategies require excellent vision, and the male's compound eyes are relatively larger than those of most bees (see photo on page 116).

ABOVE | A female *Caupolicana fulvicollis* collects nectar from a *Nolana divaricata* flower.

OPPOSITE | Nest entrances of *Caupolicana fulvicollis* honeycomb this small sandbank in coastal Chile.

There is only one species in this genus, *Diphaglossa gayi*. These large, bright orange bees can be found in south-central Chile, where they likely obtain protection from predators by looking like the local native bumble bee, *Bombus dahlbomii*, which is similarly bright orange (albeit with black hairs on the legs and ventrally, while *D. gayi* lacks black hairs altogether). The females particularly favor collecting pollen and nectar from the tubular, bright red flowers of the small Chilean Firebush tree (*Embothrium coccineum*), but can also be found on other flowers. The tongues of short-tongued bees are not very good at obtaining nectar from deep flowers,

BELOW | A female *Diphaglossa gayi*, snapped just as she was about to alight on a Chilean Firebush tree, a preferred floral host.

RIGHT | While *Diphaglossa gayi* was once thought to rely upon Chilean Firebush trees (*Embothrium coccinum*) for floral resources, detailed observations combined with studies of pollen in the scopa of females reveal that the bees are not restricted to this one plant species as a host. This bee has pollen of two different colors in its scopa.

GENUS
Diphaglossa

DISTRIBUTION
Southern half of Chile, north of 42°S

HABITAT
Relatively verdant areas, including within fairly dense woodland

CHARACTERISTICS
- Large bees with extensive orange markings
- Episternal groove absent below scrobe
- Notaulus deep
- Glossa deeply bifid
- Malar space at least two-thirds as long as compound eye

but *D. gayi* seems to compensate for this by having an unusually elongate malar area. As with all members of its eponymous subfamily, the Diphaglossinae, this species has a deeply forked glossa, which likely also aids with nectar acquisition.

Nests are deep vertical burrows that have long lateral branches that lead to single cells. The lateral branches bend upward and then downward, and end in a vertical brood cell. This pattern looks somewhat like a fermentation lock used in beer brewing and wine making, and as the relatively liquid provisions ferment, it is not impossible that it functions in a somewhat similar manner. The larvae spin cocoons, beneath which they have scooped out a space where their feces accumulate.

PARACOLLETES

This is the only genus in its tribe, with eight species. Its phylogenetic position has changed recently, initially as a result of DNA-sequence data that suggested a relationship to the Diphaglossinae (pages 118–21). However, it is also known to spin cocoons, and as (with one exception) the only other colletids that spin cocoons are diphaglossines, this shared behavior also suggests a close relationship between the two. The revised placement of this genus necessitated the reclassification of many colletids into the subfamily Neopasiphaeinae— they had all previously been considered paracolletines.

There are eight species of these moderately large bees, superficially resembling *Melitta* (page 49) or *Andrena* (page 58), but as neither of those genera occurs in Australia, which is the only place *Paracolletes* is found, there is little chance of confusing them. *Paracolletes* nests are often beneath leaf litter and it seems that females may reuse old nests.

GENUS
Paracolletes

DISTRIBUTION
Throughout most of Australia

HABITAT
Best known from sclerophyllous forest with soft, loamy soil

CHARACTERISTICS
• Colletids lacking a premental fovea

• Marginal cell bent away from wing margin toward apex

• Mandible tridentate

• Stigma small, parallel-sided

• Posterior tibial spur serrate or with slender teeth

• Jugal lobe two-thirds as long as vannal lobe

LONCHOPRIA

Bees in this genus are often beautifully colored, with bright fulvous pubescence (overlying a metallic purple integument in one undescribed species) or with nicely patterned white, gray, and black hairs against a black or metallic blue to green integument. There are 18 described species, with at least another dozen awaiting description.

The nominate subgenus is famous for having males that use their mandible to hold on to the females by their "waists." As a result, sexual selection has operated on the shape of the male mandible as well as on the details of the junction of the female meso- and metasomas. It is therefore often easy to identify species simply by looking at the shape of the male mandible.

BELOW | This female *Lonchopria zonalis* has already collected a large amount of pollen in her hind leg scopa.

GENUS
Lonchopria

DISTRIBUTION
South America, from Colombia to Patagonia; most diverse in Argentina

HABITAT
Diverse, from arid to semiarid areas, and from sea level to over 13,000 ft. (4,000 m)

CHARACTERISTICS
- Three submarginal cells
- Metatibial scopal hairs plumose and very dense, obscuring underlying integument
- Outer surface of metabasitarsus weakly concave, with sparser, shorter hairs than on metatibia
- Female posterior metatibial spur pectinate, with 4–8 teeth closely approximated basally

NEOPASIPHAE

It is somewhat unfortunate that this genus provides the name for an entire subfamily given that it is perhaps the least representative in appearance, being the only one with extensive yellow markings on the disks of the metasomal terga. There are three described species, one restricted to the eastern half of Australia and the other two (as well as three undescribed species) known only from the western half of the continent.

RIGHT | Very short, dense hair on the mesoscutum is a feature that crops up among various groups of bees. In this female *Neopasiphae mirabilis* it contrasts nicely with the yellow-banded metasoma.

BELOW | This male *Neopasiphae mirabilis* is easily recognized by the expanded hind leg (especially the pale yellow basitarsus) and flattened, bright yellow scape.

GENUS
Neopasiphae

DISTRIBUTION
Australia

HABITAT
Semiarid areas where their host plants are common

CHARACTERISTICS
- Laterally notched or interrupted yellow transverse preapical metasomal bands
- Stigma large
- Female mandible with two teeth
- Female without gradulus on S2
- Male scape and metabasitarsus swollen

One of the western species is considered to be in danger of extinction.

All species can be found feeding on flowers of the family Goodeniaceae, which often grow as temporary spring pools dry out. Females have hooked hairs on their face that collect the pollen as they enter the flowers of their hosts. In some places males spend the night in flowers, which close around them in the evening and open and release them in the morning.

As can be seen in the above image, *Neopasiphae* bees sometimes have dense short hairs on the dorsal surface of the thorax. This type of pubescence crops up irregularly in at least some species of most bee families.

CALLOMELITTA

OPPOSITE | Having a mesosoma that is paler in color than the metasoma is rather rare among the bees, but this is the case in this *Callomelitta picta*. In other members of the genus the mesosoma is dark and the metasoma red, or most of the body is entirely brown.

BELOW | This *Callomelitta* female has a coarsely sculptured integument, with lots of small pits that cover most of her body.

This is a small genus of 11 species in a subfamily all by itself. Unusually, it has been predicted that a taxonomic revision might reduce, rather than increase, the number of species. They are relatively heavily sclerotized black (or dark blue) and red or orange-brown bees. One species has been found nesting in soft wood, which is excavated by the bee using its three-toothed mandibles: in general, whenever a bee has more than two teeth, it seems to be associated with special nesting activities, such as fiber acquisition or leaf-cutting in megachilids, or chewing wood, as is the case here. Brood cells are lined with the cellophane-like material typical of colletids and are sealed with chewed soft wood.

GENUS
Callomelitta

DISTRIBUTION
Australia and New Caledonia; more abundant and diverse in the east, although one species extends west as far as Perth

HABITAT
Mostly found in the wetter parts of what is a rather arid continent

CHARACTERISTICS
- Three submarginal cells
- Mandible with a total of three teeth
- Margin between metapostnotum and propodeum generally carinate
- Apex of marginal cell on wing margin
- Stigma large
- Female pygidial plate with sides concave basally, apex narrow and more parallel-sided

This genus contains more than 500 species distributed throughout the world except Australasia, although it is less common in tropical Africa and Asia. The S-shaped second recurrent vein of the forewing is unique among all bees. *Colletes* are commonly known as cellophane or polyester bees because of their brood-cell linings, but this is a little imprecise as all members of their family make similar linings. Some North American members of the genus are among the first bees to emerge in spring, with species such as *C. inaequalis*

forming large aggregations in sandy soil. Here, males can be found in high densities where virgin females are emerging. In much of Europe and North America there are also numerous species active late in summer and early autumn. Some species have become pests due to nesting in sandstone-brick walls.

Many species have a narrow foodplant choice; for example, western European populations of *C. hederae* collect pollen and nectar from Common Ivy (*Hedera helix*), which flowers in late autumn, meaning that these bees are active later in the year than almost all others. Some species are likely economically important pollinators— for example, of tomatilloes and ground cherries (both in the genus *Physalis*). The author discovered large numbers of one tomatillo specialist in his garden for the first time in 2018, despite growing the plants there since 2005.

LEFT | A crowd of male *Colletes halophilus* swarm around a patch of ground where virgin females are emerging from the natal subterranean nest where they spent the winter.

RIGHT | Most northern hemisphere *Colletes*, such as this *C. hederae*, have orange hairs on the head and thorax, but pale apical bands on the metasoma. The coloration of species in South America is more diverse.

GENUS
Colletes

DISTRIBUTION
Worldwide except east of Wallace's Line, Iceland, Greenland, and some parts of tropical Africa and the Amazon Basin

HABITAT
Very diverse, from semiarid areas to salt marshes, and from sea level to over 10,000 ft. (3,000 m) in altitude

CHARACTERISTICS
- The S-shaped second recurrent vein is diagnostic for this genus
- Somewhat conical metasoma with a heart-shaped face
- Metapostnotum short, margined with carina, and with longitudinal carinae separating surface into pits

SCRAPTER

Scraptrinae is a colletid subfamily with a single genus: *Scrapter*. These ground-nesting bees are restricted to Africa and are most abundant and speciose in the Cape Floristic Region; only one species is known outside of southern Africa. Their appearance is very variable: some are small, shiny, black, and relatively bald, while others are larger and abundantly hairy. It seems probable that the genus might deserve division into multiple genera.

Molecular phylogenetic analyses suggest that these bees are most closely related to the Australian Euryglossinae, and the smaller species are informally referred to as euryglossiform *Scrapter*, suggesting that there are at least superficial morphological similarities between them. There are 68 described species, with doubtless quite a few more awaiting discovery.

GENUS
Scrapter

DISTRIBUTION
Southern Africa; an earlier record from Kenya is considered dubious

HABITAT
Semiarid areas

CHARACTERISTICS
- Colletidae with a premental fovea that has shiny lateral ridges converging apically
- Two submarginal cells
- Galeal comb with three or four bristles
- Keirotrichiate area with long hairs

EURYGLOSSA

The entire subfamily Euryglossinae is restricted to Australasia and many of its species collect pollen, primarily from eucalypts. While many species are minute, some are of a more average size for a bee; the 37 species of *Euryglossa* are among the largest and most robust of the subfamily. The females range in color from pale green, through mostly red to black and entirely metallic blue. Males are mostly black and are very different in appearance from the females, such that it is difficult to believe they are alternate sexes of the same species. Their cylindrical body shape would suggest that they might nest in tunnels in solid wood, but those nests that have been found are in the ground.

LEFT | The expanded hind femur and tibia of this male *Scrapter heterodoxus* can easily be seen in this image.

RIGHT | Sexual dimorphism is quite well developed in *Euryglossa*, even in the absence of unusual secondary sexual characteristics, as can be seen in this courting pair of *E. adelaidae*.

GENUS
Euryglossa

DISTRIBUTION
Throughout Australia

HABITAT
Areas with eucalyptus trees or other Myrtaceae

CHARACTERISTICS
- Broad, weakly concave glossal apex
- Females with swollen sides to mesoscutum, which is also parallel-sided anterior to tegulae
- Head quadrate
- Males with long antennae
- Posterior margin of first submarginal cell sinuate
- T1 much broader than long

PACHYPROSOPIS

There are 23 described species in this genus, varying considerably in size and color. Some are mostly yellow or orange, or pale with black markings, while others are mostly metallic blue. Although they are often remarkably unbee-like bees, their morphology is also variable, some with relatively square heads and others with acute processes on the gena. Larger individuals of one species have a pair of curved horns on the clypeus; the smaller the individual, the smaller the horn, with the smallest individuals merely having a small angulation. It is not known what this feature is for.

Like many Australian bees, *Pachyprosopis* forage primarily on flowers of eucalypts and their relatives. Nest sites have been found in an old tree stump and sawdust-like debris in a tree hollow.

GENUS
Pachyprosopis

DISTRIBUTION
Throughout Australia

HABITAT
Areas with eucalyptus trees or other Myrtaceae

CHARACTERISTICS
- Broad, weakly concave glossal apex
- Second submarginal cell with anterior margin strongly angled toward costal margin of wing
- Female compound eye extending medially below, so that anterior mandibular articulation is near midline of eye

CHILICOLA

This is a diverse genus of mostly small, twig-nesting South American bees, in which the number of described species is likely less than half the number actually held in museum collections. Some species have elongate heads, although they are not as long as in some species of the related *Geodiscelis* (page 9) or *Xeromelissa* (page 137). Some are unusually narrow and rather unbee-like. Others have enormously modified hind legs in the males, with swollen femora and tibiae expanded into a wide range of shapes and sporting variously developed spines and ridges. None of the long-headed species have such unusual hind legs, perhaps suggesting that it is difficult to have extreme morphologies in two different parts of the body at the same time. Most species are black, some have red markings, and there is one small undescribed species that is almost entirely brick red. Unusually, females of most species have a corbicula on S2, where the scopal hairs are posteromedially oriented but absent medially.

This is the only genus in the entire family Colletidae for which fossils are known: two species have been found in Dominican amber. Their morphology is so well preserved that they can clearly be placed within phylogenies of *Chilicola* species.

LEFT | This male *Chilicola chalcidiformis* sports brightly colored and rather strongly expanded metatibiae. These likely provide both visual and physical sensory experiences, the first perhaps aimed mostly at other males and the latter perhaps at females.

RIGHT | This is an undescribed species of the *Chilicola* subgenus *Hylaeosoma*. Such morphological diversity (compare this image to the one to the left) might suggest the genus needs to be divided. This individual is an unusually narrow member of the genus.

GENUS
Chilicola

DISTRIBUTION
From Mexico to the southern tip of South America; most diverse in Argentina, with one species in Saint Vincent

HABITAT
Diverse, from tropical islands and rain forest to arid deserts, from sea level to over 13,000 ft. (4,000 m) in altitude in the Andes

CHARACTERISTICS
- Scopa as well developed on S2 as on the hind leg
- Premental fovea present
- Pale bands absent, metasoma sometimes red or red-marked
- Maxillary palpomeres similar in size and shape
- Strong epistomal lobe absent
- Metapostnotum usually longer than metanotum

GEODISCELIS

For a group of only six species, *Geodiscelis* exhibit a remarkable range of variation. Two species have heads that are approximately round, while the others have elongate heads, one of them spectacularly so, and one species has males with perhaps the most extremely modified hind leg of any bee. The first species of this genus was discovered in the late 1990s, indicating that there is still ample opportunity to find remarkable bees that nobody has seen before. That said, as the second-known species was found near the ancient Inca Trail in northern Chile, it is possible that earlier inhabitants of the land noticed the strange, superficially mosquito-like bees flying around the last patches of vegetation before the road descends from the mountains into the absolute desert.

BELOW | *Geodiscelis longiceps* is known from a very small area on the edge of the "absolute desert" of northern Chile, part of the Atacama Desert, where it obtains nectar from flowers with very deep nectaries.

GENUS
Geodiscelis

DISTRIBUTION
Northwestern Argentina, northern Chile, and southern Peru

HABITAT
Semiarid to extremely arid deserts

CHARACTERISTICS
- Maxillary palpomeres all similar in size and in their cylindrical shape
- Pale markings on metasomal terga
- Dense patches of short, broad, appressed hairs
- Premental fovea present but small
- Strong epistomal lobe present

XEROMELISSA

This is a diverse genus with 22 described species but at least another 40 awaiting formal description. They are primarily oligolectic bees. Some species have enormously elongate heads, while others have heads that are round, and some males have strongly modified hind legs, while others have very slender legs like the females. The long-headed species usually also have exceptionally long mouthparts, which they use to access nectar in the deep nectaries of their desert flower hosts. One undescribed species has a brush of coarse bristles on the mouthparts that seem to scrape pollen out from the small flowers it specializes on.

Some species nest in dead twigs that have been hollowed out by beetles, while others make nests in volcanic ash. Most species have yellow and orange markings on the metasoma, which in combination with the patches of white hairs on their otherwise black bodies makes it hard to see them when they land on the desert sand.

LEFT | *Xeromelissa sororitatis* is unusual in that it nests in the cavities of volcanic pumice.

GENUS
Xeromelissa

DISTRIBUTION
Southern South America, from Peru to Patagonia

HABITAT
Semiarid to extremely arid areas. Some species require plant stems hollowed out by beetles for their nests, others soft sandy soil or volcanic ash

CHARACTERISTICS
- Basal maxillary palpomeres larger than apical ones; apical palpomeres very small, rarely absent
- Pale markings on metasomal terga
- Premental fovea present
- Strong epistomal lobe present

A genus of more than 750 species, these bees are unusually wasp-like in that they carry pollen and nectar back to the nest inside their digestive system rather than on an external scopa. Relatively hairless, they are commonly known as masked bees because of the usually white to yellow face of males, and the triangular area of this color on each side of the face in females. A few species have red markings on the metasoma, but it is in Australia and New Guinea that their morphology

OPPOSITE | Female *Hylaeus* usually have a pair of pale triangular markings on the face.

BELOW | Bees of the genus *Hylaeus* are commonly called masked bees because of the pale face of the males, as can be seen in this *H. nigritus* male.

GENUS
Hylaeus

DISTRIBUTION
Almost worldwide, even on relatively isolated islands

HABITAT
Diverse, from arid to moist areas

CHARACTERISTICS
- Scopa absent
- Body with short, sparse hair
- Supraclypeal area usually abruptly raised
- Premental fovea present
- Both recurrent veins join second submarginal cell

and color patterns become particularly diverse, with some spectacularly bright blue, green, or purple metallic species, others that are almost entirely yellow.

Most species of this genus nest in pithy stems or other hollowed-out plant material, including vacated plant galls, although a few nest in the ground. Quite a few species have been accidentally introduced to non-native continents, undoubtedly facilitated by their habits of nesting in dried plant material. Some European species have been found in North and South America, and one Australian species has been found in Chile. Dried plant material also floats well, and this might be the way some species have found their way to isolated islands such as Hawaii, where one particular subgenus has radiated, even evolving a kleptoparasitic lineage of five species. Seven of Hawaii's *Hylaeus* were the first bees listed as endangered in the USA.

MEROGLOSSA

OPPOSITE | Like some others in the subfamily Hylaeinae, some *Meroglossa* species will take to nesting in artificial bee hotels, as can be seen with this *Meroglossa rubricata*, photographed in Perth, Australia.

BELOW | Some *Meroglossa* species come with different color-coded body parts, making them ideal for teaching bee morphology.

This is an unusual genus of colletid bees owing to the fact that males have a pointed glossa that is entirely atypical for the family. The concave glossa of Colletidae was thought to indicate that they were the earliest-diverging bee family, as the wasps from which bees arose have a similarly shaped glossa. However, detailed analyses of the ultrastructure of the tongues of bees and wasps indicate that the situation is more complex.

GENUS
Meroglossa

DISTRIBUTION
Australia except Tasmania, and less diverse in the west

HABITAT
Most habitats within its range

CHARACTERISTICS
- Glossa concave in females, pointed in males
- Face with depressions, often elaborate in form
- T2 fovea small, round to absent
- T2 gradulus bowed posteriorly medially
- Hind tibia with one or two apical spines

Finding some genera of colletid with a pointed glossa further casts suspicion on the colletidae-basal hypothesis, which was finally put to rest with the advent of large-scale DNA-sequence data.

Meroglossa is an endemic Australian genus of 20 species. They nest in wood or pithy stems and can be attracted to so-called "bee hotels." Many species are unusually colored, with large areas of red or brown, while others are black or dark blue with pale markings.

MEGACHILIDAE

Most megachilids have an obvious scopa on the underside of the metasoma rather than on the hind legs, a few collect pollen on most of their body surface, and the cuckoo species have no scopa at all. The combination of having a long-tongued bee morphology and a labrum that is widest at the base separates them from all other bees.

There are four generally accepted subfamilies in the Megachilidae, three with relatively few species. Fideliinae includes 19 species of ground-nesting bees found in Chile and southern Africa, with one species each in Peru and Morocco. They are the only megachilids with three, rather than two, submarginal cells. Pararhophitinae contains a single genus of three small species restricted to arid areas extending from Morocco to northwestern India. The combination of extensive pale markings on the integument and the lack of any subapical teeth on female mandibles suffices to separate them from other megachilids. They make shallow nests in the ground. Lithurginae has 64 species in five genera. They can be differentiated from other megachilids by the combination of two submarginal cells, a lack of pale markings, and the presence of pygidial plates in both sexes, albeit reduced to a small spine in females.

They exist on all continents that have bees and have dispersed to many islands. They excavate nests in wood and similar substrates, including dead cacti and mammal droppings.

In contrast to the above, the subfamily Megachilinae contains more than 4,000 species divided into a large, but ever-changing, number of genera. The combination of two submarginal cells, at least two subapical teeth to the mandible, and the absence of a pygidial plate serves to identify them. This subfamily contains the most economically important solitary bees, including the Alfalfa Leafcutter Bee (*Megachile rotundata*) and a range of orchard bees. They have the broadest range of nesting behaviors of any bee subfamily, with nests in preformed cavities (such as holes in wood, tunnels in pithy stems, abandoned bee or wasp nests, and even patio furniture and other human-made hollows), constructed on the surface of stones or plants, or excavated in the soil. They also bring back a wide variety of materials to the nest to line their brood cells, including cut or chewed leaves or petals, plant or animal hairs, mud, resin, and even pieces of plastic or tile caulking.

LEFT | *Trachusa larreae* feeding on the plant that gives it its species name: *Larrea tridentata* or Creosote Bush.

Most individuals of this genus of five species present in collections are covered in long grayish hairs, but if you find one that has only just become an adult, these hairs are pale orange: the color rapidly fades in the bright sunshine of the deserts where they live. At least three species collect pollen only from Chilean bell flowers (*Nolana* spp.). The bees have a long glossa to reach the nectar in the deep flowers, with one exception: *Neofidelia profuga* has a broad range of floral hosts and a comparatively short tongue, and lives in the southern parts of the Atacama Desert, where rainfall and floral diversity are both higher.

One species is known only from a fog oasis on the Chilean coast; such oases form only under very specific conditions, where damp air, moistened by the cold Humboldt Current, blows over a promontory that extends into the ocean. The moisture condenses into fog as it rises, and if there is a bay surrounded by high hills on the leeward side, it becomes trapped, resulting in an almost continuously wet environment even though it almost never rains in the region. Elsewhere, the fully grown larvae must detect the increased moisture after rare rainfall from within their brood cells.

LEFT | This male *Neofidelia* looks like it would be a good pollinator even though it is not actually collecting pollen on purpose.

RIGHT | This male *Neofidelia profuga* is sitting on a *Nolana* flower waiting for a female to impress with his massively expanded metafemora and other accoutrements.

GENUS
Neofidelia

DISTRIBUTION
Northern Chile and southern Peru west of the Andes

HABITAT
Areas of low rainfall. Populations of two species are seemingly maintained by coastal fog. One more occurs in the southern Atacama Desert, where rainfall is more predictable, and another at high altitudes where there is sufficient rain for bee activity in some years only

CHARACTERISTICS
- Three submarginal cells
- Males have swollen hind femora and claw-like metabasitarsi
- Females have very long hairs on the dorsal surface of the metatibia

XENOFIDELIA

Sometimes, even with decades of experience, a melittologist finds something that is completely surprising and initially impossible to categorize. These are exciting moments. Only after looking at the single known specimen of this genus under the microscope for a long time did it become clear to the author that it was related to *Neofidelia* (page 146), but with substantial differences—such as some aspects of the mouthparts looking more like those of a short-tongued bee. Despite months of trapping around the locality where this bee was caught, no more individuals have been found.

DNA sequences obtained from this specimen confirmed that its closest relative is *Neofidelia*, from which it diverged more than 30 million years ago. This specimen is the sole known representative of its lineage for more than 30 million years of evolution!

BELOW | This is the only specimen of the genus *Xenofidelia* ever found, despite considerable searches.

GENUS
Xenofidelia

DISTRIBUTION
Known only from near Mamiña in northern Chile

HABITAT
An arid area at 8, 200 ft. (2,500 m) in the northern Atacama Desert, where rain—if there is any—falls primarily in summer

CHARACTERISTICS
- Three submarginal cells
- Glossal rod absent

There are only three species in this little
subfamily of bees, which is restricted to arid
areas from Morocco to northwest India. The
females are mostly pale in color, while the males are
darker but still with extensive pale markings.
Females collect pollen over a comparatively large
surface of their bodies, legs, mesosoma, and to
some extent even the metasomal dorsum, not just
on the underside of the metasoma, as in most other
members of the family Megachilidae.

The bees make shallow nests in the ground,
with brood cells isolated at the end of lateral
branches. The females construct a space at
the end of the brood cell that contains
sand, perhaps mixed with nectar. The
fully grown larva eats this pellet and
incorporates the sand into the inner
lining of the cocoon when it
defecates.

BELOW | It is unusual for a bee
to have an almost entirely yellow
exoskeleton, as in this *Pararhophites
orobinus* female. The males have more
black markings.

GENUS
Pararhophites

DISTRIBUTION
Morocco to northwest India

HABITAT
Deserts

CHARACTERISTICS
- Two submarginal cells
- Mandible lacking subapical teeth
- Extensive yellow markings

OPPOSITE | Male *Lithurgus chrysurus* are more uniformly covered in yellow-orange hair than are the females.

BELOW | This female *Lithurgus chrysurus* demonstrates the meaning of its species name, which can be translated as "yellow tail," although in reality her T6 is covered in hairs more orange in color.

Given the importance of bees to agricultural production (even meat and dairy through the pollination of alfalfa) and the pollination of wild plants, it seems incongruous to suggest that a bee species might be a pest (although those allergic to bee venom might consider all stinging insects to be a hazard). However, one species of the genus *Lithurgus*, *L. chrysurus*, could be considered a pest because the females chew into even quite hard

GENUS
Lithurgus

DISTRIBUTION
Worldwide, including many small islands in warm climates; absent from more northerly regions of Eurasia and central Australia. Not native to North or South America, although one species has been introduced to eastern North America, another likely to Brazil, and a third to Hawaii

HABITAT
Diverse, including arid areas, rain forest, and tropical islands

CHARACTERISTICS
- Mid- and hind tibiae tuberculate
- Pygidial plate narrow and dorsally concave, spine-like in females, peg-like in males
- Female mandible with lower tooth clearly shorter than middle tooth

- Arolia absent or minute in both sexes
- Female face with prominence on upper part of clypeus and sometimes on supraclypeal area
- Metapleuron very narrow for ventral half
- F1 twice as long as wide

wood and have been found causing damage to roof shingles as well as the underlying structural beams.

Bees of this genus as a whole nest in wood—often soft wood that is easier for them to excavate. The brood cells are not lined in any way and may, or may not, be separated by wooden partitions. The lack of partitions between adjacent pollen balls is a rare phenomenon.

TRICHOTHURGUS

A genus of 14 species of mostly rather large bees with elongate pretarsi, especially in the males, which leads to the potential common name of great grapplers. They nest in soft plant material such as rotting wood, and one species has been found nesting in horse dung. These are mostly specialist bees, with some species concentrating on cactus flowers—the males can be seen in the flowers awaiting the arrival of females. Others have been seen foraging on flowers that are quite small and with thin stems for the size of the bee visiting, with the result that the flower stalk bends, sometimes to the ground, as the bee alights, and then springs back up again as it leaves.

BELOW | A *Trichothurgus dubius* female feeding on a *Centaurea chilensis* flower.

GENUS
Trichothurgus

DISTRIBUTION
Western South America, from Peru to southern Argentina, including Patagonia

HABITAT
Arid to semiarid areas, from desert to dry scrub

CHARACTERISTICS
- Mid- and hind tibiae tuberculate
- Pygidial plate narrow and dorsally concave, spine-like in female, peg-like in male
- Tarsal claws of females lacking teeth
- Arolia absent or minute in both sexes
- Labrum longer than clypeus
- Lower mandibular tooth at least as long as middle tooth

OCHRERIADES

This is an interesting genus and tribe of only two species, one known from a single specimen from Namibia. Despite its few species, *Ochreriades* has caused some headaches for bee taxonomists because its members seem to possess a mixture of features of the Anthidiini and Osmiini. Perhaps it is a general phenomenon that when there is controversy over placement of a taxon, it deserves its own higher-level category, as is now the case for the two species in the Ochreriadini.

These are narrow cylindrical bees, and like other similarly shaped bees, they nest in burrows in wood that have already been excavated by other creatures. There are 1–5 brood cells per nest, separated by mud partitions, and the entrance is closed by pebbles embedded in mud.

RIGHT | This long, narrow bee— *Ochreriades fasciatus*—obtains pollen only from plants of the mint family (Lamiaceae).

GENUS
Ochreriades

DISTRIBUTION
Disjunct, known only from the Middle East and Namibia

HABITAT
Semiarid areas where their host plants and suitable nest sites are found

CHARACTERISTICS
• Arolia present

• Stigma more than twice as long as broad

• With pale markings

• Mouthparts very long

• Body long, narrow, and cylindrical; the long, elevated pronotum increases the cylindrical appearance

• Female labrum lacking long erect hairs apically

DIOXYS

DIOXYS

BELOW | As with all cuckoo bees, this *Dioxys cincta* female lacks a scopa. As the bee is a megachilid, we would expect to see the scopa on the underside of the metasoma, but it is fairly easy to see that this feature is absent here.

Most cuckoo bees have long stings (as in *Osiris*; page 186), and strongly sculptured bodies with deep pits and strong ridges to deflect the weapons of any host that might return home while the thieves are performing their kleptoparasitic acts. This genus of 18 species, along with the seven other genera in its tribe, is unusual in that while possessing strong armor, its members have completely reduced stings—even more reduced than in the stingless honey bees (which have other modes of defense including biting).

These cuckoo bees are not as specific about their hosts as are species of most kleptoparasitic genera. At least one has been found to attack nests of Anthidiini, Osmiini, and Megachilini. The first three instars of the larva retain the sickle-shaped mandibles, which are capable of slashing the eggs or small larvae of hosts or other cuckoo bee individuals.

GENUS
Dioxys

DISTRIBUTION
Western North America, from southern Canada to northern Mexico, and circum-Mediterranean, extending to central Asia

HABITAT
Diverse, from arid areas to forests and meadows where host populations are healthy

CHARACTERISTICS
- Kleptoparasitic Megachilinae
- Metanotum with a median spine
- Scutellum with lateral processes without a carina between them
- Labrum at least as long as mandible and parallel-sided
- Preoccipital carina strong
- Labrum without transverse basal carina

ANTHIDIELLUM

The small bees in this genus are rather wide for their length. They include 63 species with a relatively diverse suite of morphologies. A small group of tropical Asian species have strongly colored wings with brown and orange markings, and superficially look more like beetles than bees.

The bees make nests of resin on the surface of plants or rocks. Sometimes only one brood cell is formed in each nest, and at other times small groups of cells. Some species typically decorate the outside of the nest with pieces of gravel. The tropical Asian species mentioned above nest in the ground, but they make their brood cells and a turret at the nest entrance out of resin.

BELOW | As with most anthidiines, this *Anthidiellum notatum* has yellow markings on its head, mesosoma, and metasoma.

GENUS
Anthidiellum

DISTRIBUTION
Worldwide, but absent from the southern half of South America, most of Australia, and the far north

HABITAT
Diverse, from cold temperate to tropical areas, and from arid scrub to rain forest

CHARACTERISTICS
- Arolia present
- Propodeum with a fovea margined by a carina behind spiracle
- Omaulus carinate to lamellate and attaining at least mid-depth of thorax
- Metapostnotum with pits at least laterally
- Scutellum produced posteriorly, truncate or medially emarginate in dorsal view
- Subantennal sulci usually strongly arcuate

ANTHIDIUM

The life of *Anthidium* larvae seems rather luxurious: they grow up in a tiny pillow—a brood cell inside a ball of plant fibers—provisioned with all the food they need. The females shave "hairs" from plant leaves with their long, multitoothed mandibles, forming a ball of fluff to take back to the nest. Once the female has constructed the pillow, she hollows it out,

constructs a pollen ball, lays an egg, and plugs the cell entrance with more plant fibers. A row of such brood cells is constructed inside a cavity in wood or sometimes in a burrow in the ground. The behavior of fabricating brood cells in this manner gives the bees their common name: wool carder bees.

Male wool carder bees can be fiercely territorial, chasing off all-comers to their territory, which is usually a small patch of plants of the type that the females will visit for resources (plant hairs, pollen, and/or nectar). They chase away all intruders other than conspecific females they attempt to mate with. They inflict aerial head butts on large bumble bees and will crush persistent honey bee workers between the long teeth at the end of their metasoma and underside of their mesosoma.

There are 188 species in the genus. One European species has been accidentally transported from Europe almost worldwide—to South America, North America, New Zealand, and Australia.

GENUS
Anthidium

DISTRIBUTION
Worldwide west of Wallace's Line, although introduced to both Australia and New Zealand

HABITAT
Diverse, from equatorial rain forests to cold temperate meadows, and from sea level to well over 13,000 ft. (4,000 m) in altitude

CHARACTERISTICS
• Megachilinae with extensive yellow markings
• Female mandible with at least five sharp teeth

- T5 apical impressed area broadly triangular, usually densely punctate, punctures smaller than on rest of tergum
- Arolia absent
- Stigma and prestigma short

ABOVE | A female Wool Carder Bee (*Anthidium manicatum*) is carding the hairs on the underside of a leaf and making them into a ball. She will carry this back to the nest to make a pillow, inside which her offspring will develop.

OPPOSITE | *Anthidium* are attractive large bees with interesting behaviors that can be followed with the use of binoculars from the comfort of a lawn chair.

RHODANTHIDIUM

At least some of the 14 species of this genus nest in snail shells and also use them for shelter during inclement weather and at night. Like most Anthidiini, males are larger than females and will often defend a snail shell that is being provisioned by a female and fight with other males over the tiny territory. The male will mate with the female frequently as she returns with resources for the offspring. After completing provisioning for one or two offspring, the female fills the rest of the shell with shell fragments that are held together with a shiny mixture of sand and salivary secretions. She then transports the shell to a suitable location and buries it in the sand; this can take several days.

BELOW | A female *Rhodanthidium sticticum* has completed a pollen ball or two, laid an egg on each, and is now closing the entrance to the snail shell she chose as a nest site.

GENUS
Rhodanthidium

DISTRIBUTION
Circum-Mediterranean and central Asia

HABITAT
Areas with soft calcareous soil and healthy snail populations

CHARACTERISTICS
- Arolia present
- Male T7 with three or five spines
- Female S6 unmodified
- Vein cu-v of hindwing less than half as long as second abscissa of M+Cu
- Body with yellow or reddish markings

SERAPISTA

While most anthidiines have extensive pale markings (usually yellow, as in the *Anthidium* on page 156), *Serapista* are black with white patches of fur. Some species have red legs and the males may have yellow on the clypeus and mandibles. They have interestingly serrate sides to the metasoma. There are six species in this genus.

Like *Anthidium*, these bees also collect hair-like fibers from plants, but unlike members of that genus, they attach the large ball of fluff they make to a plant stem. Sometimes animal hairs, or even bird feathers, are incorporated into the nest, inside which brood cells are formed and provisioned with pollen and nectar.

BELOW | The entirely black body with patches of white hairs is an unusual phenotype among bees of the Anthidiini.

GENUS
Serapista

DISTRIBUTION
Sub-Saharan Africa

HABITAT
Diverse, from semiarid areas to rain forest

CHARACTERISTICS
- Metasomal terga with lateral spines
- T6 of female and T7 of male with longitudinal median carina
- Pronotal lobe and scutellum lamellate
- Female mandible with numerous sharp teeth
- First recurrent vein goes to the first submarginal cell
- Apical impressed areas of metasomal terga not notably broader medially

TRACHUSA

This is a genus of bees that, at least superficially, appear very diverse. Some would easily be mistaken for an *Anthidium* based on overall color pattern and shape (see page 144), while others look more like a *Megachile* because they lack pale markings entirely (except for the clypeus of males). Still others look like bumble bees—they are medium-sized, rather rotund, and quite hairy. There are 64 species and bee taxonomists demonstrate awareness of the phenotypic diversity of these by dividing them into no less than 10 subgenera, six of which are found only in the eastern hemisphere and four only in North America, including Mexico. None has yet been found in Canada, although some species are getting closer (being known from Minnesota and Michigan, for example) and might move further northward as the climate warms.

Unlike most other anthidiines, *Trachusa* nest in burrows in the ground. They excavate these themselves and line the brood cells with either just resin or a mixture of resin and plant material.

GENUS
Trachusa

DISTRIBUTION
Throughout most of the eastern hemisphere, as far north as Finland, as far east as eastern Russia (although generally absent from Siberia), and through much of Africa and tropical Asia to Borneo; absent from Madagascar and Australasia. In the western hemisphere, from south of the US–Canada border to southern Mexico

HABITAT
Diverse, from semiarid scrub to tropical forest wherever suitable host plants and nesting soil can be found

CHARACTERISTICS
- Stigma and prestigma short
- Vein cu-v of the hindwing oblique and usually almost half the length of the second abscissa of M+Cu

- Carinae and lamellae restricted to, at most, the pronotal lobe and omaulus
- Lateral ocelli closer to compound eye than preoccipital margin or equidistant to it
- Midtibia relatively broad, with anterior and posterior margins convex
- Male T7 small and oriented anteriorly, curled under

ABOVE | As can be seen here, *Trachusa cordaticeps* is one of the more attractively patterned species of bee in North America.

OPPOSITE | Comparing this image of *Trachusa manni* with the one above and on page 144 demonstrates the diversity of integumental color patterns found within the genus *Trachusa*.

COELIOXYS

OPPOSITE | Male *Coelioxys* have unusual tail ends, with multiple angles and processes as in this *C. conoidea.*

BELOW | The sharp, pointed apex of this *Coelioxys elongata* shows clearly that it is a female. The dagger-like tail is used to gain access to the host's brood cells.

These bees are remarkable for the sharp tail end of the metasoma in females. As most species in the genus lay eggs inside the nests of *Megachile* (leafcutter or resin bees; page 166), the sharp end of the metasoma serves to deposit the egg into the pollen ball or into the cell lining while the host female is out foraging, or into the cell lining after the brood cell is closed. It is not until the first-instar larva has molted into the second stage that it develops mandibles capable of killing the host egg and eggs of conspecifics deposited in the same brood cell. The males also have unusual ends to their abdomens, usually with four large teeth. Another uncommon feature among bees that is found in most species of *Coelioxys* is long hairs on the compound eyes (see also *Caenohalictus*, page 102).

There are more than 470 species of this cuckoo bee genus, making it the second-most speciose genus of cuckoo bee after *Nomada* (page 194).

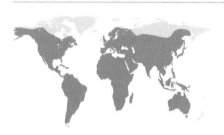

GENUS
Coelioxys

DISTRIBUTION
Worldwide, including most islands, but absent from Antarctica, Greenland, Iceland, the Arctic and subantarctic islands, and New Zealand

HABITAT
As diverse as that of their hosts— mostly *Megachile*—except not as common in the far north

CHARACTERISTICS
- Arolia absent
- Eyes usually hairy
- Transverse depressions behind the gradulus on at least T2 and T3
- Female metasomal apex acute
- Male metasomal apex with spines

GRONOCERAS

This is a genus of 12 species from sub-Saharan Africa. An old specimen from Jamaica is considered to be either a labeling error or an introduction that has not persisted. Until recently *Gronoceras* was considered to be a subgenus of *Megachile*. However, a kleptoparasitic lineage (see *Coelioxys*, page 162) diverged after the separation of *Gronoceras* and before the divergence of true *Megachile*. Consequently, *Gronoceras* was one of several groups raised to genus level.

These are large, impressive bees with patches of gray hairs or a striking red-and-black color pattern. The antennae of males of some species are expanded and appear hollowed out ventrally, giving rise to their generic name, which means "eaten out" or "hollow" horn. A nest of this genus is shown on page 27.

ABOVE | Some bees strike quite distinct poses as they feed from flowers, as with this female *Gronoceras felina*.

RIGHT | This male *Gronoceras felina* is perched on a twig, from where he surveys his surroundings looking for a potential mate.

GENUS
Gronoceras

DISTRIBUTION
Sub-Saharan Africa and the islands between the east African coast and Madagascar

HABITAT
From semiarid areas to tropical rain forest

CHARACTERISTICS
- No arolia
- No pale integumental bands
- Four mandibular teeth
- Mandible dull
- Foretibia with three distinct teeth separated by shiny, largely hairless areas
- Male with two long teeth or lobes on a T6 that is covered with long hairs

165

MEGACHILE

GENUS
Megachile

DISTRIBUTION
Worldwide, including most islands,
but absent from Antarctica, Greenland,
Iceland, and the Arctic and
subantarctic islands

HABITAT
Found in just about every terrestrial
habitat

CHARACTERISTICS
- Lacking pale integumental markings
- Stigma twice as long as broad
- Arolia absent
- Claws lacking teeth
- Mandibles with several teeth,
  often with cutting edges between
  adjacent teeth

This genus includes almost 1,500 species, even with the recent separation of some species to other genera. Some researchers wish to divide it even more finely, separating it into a couple of dozen extra genera.

As these bees usually line their brood cells with plant material such as leaves or resins, they are commonly called leafcutter or resin bees. But some have been recorded using pieces of plastic bag or kitchen tile caulking in place of more natural materials—perhaps one more symptom of the Anthropocene. When you see almost perfectly semicircular excisions to the leaves of your prize rose bush or other garden plants, the culprit is almost certainly a leafcutter bee. If you are patient, you might see one cut the leaf and then launch into flight with the piece held beneath its body. Most species nest in pithy stems or other naturally occurring holes (including the metal tubes used to make patio furniture), while others nest in the ground, but all use some form of extraneous material to line their brood cells. One species is economically important for the beef and dairy industry: the Alfalfa Leafcutter Bee (*Megachile rotundata*) excels at pollinating alfalfa (a crop poorly served by honey bees), which is a winter fodder crop for cattle. Some leafcutter bees choose leaves that have antimicrobial properties.

HERIADES

BELOW | A female *Heriades truncorum* carrying pollen on her metasomal scopa back to her nest, which is in a bee hotel. The mud nest closures can be seen at the entrances of adjacent nests.

This is a diverse, relatively large genus with 135 species of mostly small, strongly punctured osmiine bees. They nest in preformed cavities in wood or stems, and most species use resin to separate their brood cells. As a result, they are commonly found nesting in holes in the cement between bricks in walls or in wooden garden furniture, posts, or fences. *Heriades* can be found on many islands within their geographic range; as is the case with other bee genera that often nest in cavities, they may have dispersed inside floating logs.

While the metasomal sterna in males is strongly modified in many Megachilinae genera, in *Heriades* most sterna are considerably reduced and largely membranous. Thus, only two sterna can easily be seen, but even for that the metasoma has to be uncurled as the entire male metasoma is generally curled ventrally and even T7 is considerably reduced, largely membranous, and hidden by T6.

GENUS
Heriades

DISTRIBUTION
Worldwide, except east of Wallace's Line and only in Colombia in South America

HABITAT
Diverse, from arid areas to rain forest

CHARACTERISTICS
- Without pale markings
- Long-tongued bees but mouthparts not reaching front coxae
- Stigma at least twice as long as wide
- Arolia present
- Scutellum flat
- Clypeus without median carina
- Male T7 hidden by T6
- Male S2 and S3 notched laterally
- Female hypostomal area unmodified

A paper published by Samuel Boff in 2022 documented the courtship behavior of *H. truncorum*, which involves the male vibrating his mesosoma and fanning his wings and the couple "dancing" in a lateral rocking motion. Wing-fanning seemed to assist the male's physical stability during copulation. Boff also demonstrated that, if a female had a choice of mates, she preferred the larger male.

ABOVE | This male *Heriades truncorum* sits atop a flower on the lookout for a mate. It is just possible to detect the recurved metasoma in this side view.

There are 378 described species of *Hoplitis*. They are found throughout the northern hemisphere, Africa, and Asia, reaching as far southeast as Borneo. Phylogenetic analyses of the genus indicate that they originated as ground-nesting bees in dry areas of the Palearctic but dispersed to their current, almost worldwide distribution. Those groups that colonized colder and wetter areas are those that nest in wooden or pithy plant materials.

With the exception of a few brilliantly metallic (and beautiful) North American species, they are all dark blackish, although their hair may be anything from black, through orange to whitish.

The genus can be defined as those Osmiini in which the males have large, somewhat membranous flaps on S6. But not all species that seem otherwise clearly to belong to the genus based on combinations of other characteristics possess this feature and in some of them the flaps are small. This is a common pattern in evolution: a feature might arise just once, but once present it might be lost multiple times.

As might be expected in a speciose genus of megachilid, *Hoplitis* nesting biology is diverse.

LEFT | *Hoplitis biscutellae* favors flowers of the Creosote Bush (*Larrea tridentata*), which can be an abundant resource in parts of the arid American southwest.

GENUS
Hoplitis

DISTRIBUTION
Worldwide, but absent from Australasia, Madagascar, and in the western hemisphere south of Mexico

HABITAT
Diverse, from arid areas to the Taiga Plains of Canada's Northwest Territories and the Congolese rain forest

CHARACTERISTICS
- Lacks pale markings
- Stigma at least twice as long as wide
- Arolia present
- Female labrum with marginal fringe
- Parapsidal line elongate
- Male T6 usually with lateral tooth and/or with tridentate T7; male S6 usually with two pale flaps

In fact, the genus uses all solitary bee nesting sites: some dig burrows in the ground, some chew burrows in pithy stems, some nest in naturally occurring cavities (including abandoned wasp nests and snail shells), and others build nests above ground. They use a wide range of materials to construct their brood cells, including resin, chewed leaves, flower petals, gravel, and mud.

ABOVE | This *Hoplitis* male is one of the more brightly orange-haired species of the genus.

171

OSMIA

Of all bees, this genus probably has the most diverse range of nest sites. While most nest in pithy stems or holes in wood, some nest immediately beneath a stone, some choose empty snail shells, and others nest in burrows in the ground. One species was accused of causing a plane crash after parts of a nest were found in the downed aircraft's fuel line, but it was subsequently discovered that the nest had been constructed after the crash rather than causing it. Some of the species that nest in snail shells go to great lengths to keep natural enemies away from their offspring. After making a pollen ball and laying an egg in the innermost spiral of the shell, they fill the rest of the space with gravel. Some species will even hide the snail shell under a pile of cut dead stems, while

GENUS
Osmia

DISTRIBUTION
Throughout the northern hemisphere, south to Central America and subtropical Asia, and introduced to New Zealand; absent from sub-Saharan Africa and Australia

HABITAT
Diverse, from the Arctic to the tropics, and from wooded areas to open fields and arid scrub

CHARACTERISTICS
- Usually with dull metallic integument
- Parapsidal line round to oval
- Arolia present
- Stigma twice as long as broad

ABOVE LEFT | A female *Osmia cornuta* gets ready to leave her nest on a foraging trip. The two horns that give this species its name (*cornuta* means "horned") can be seen on the front of the face.

ABOVE | Bee hotels usually result in unnaturally dense aggregations of nesting sites for cavity-nesting bees. Here we see both male and female *Osmia bicornis* at such a bee hotel. Different nest-entrance plugging materials have been used by different females.

others will seal the nest entrance with chewed leaf pieces or mud.

There are 350 described species in this genus, which is found throughout the northern hemisphere. Most are dull metallic in color, although some are entirely blackish and a few have a red metasoma.

APIDAE

This family includes the most commonly recognized bees in the world—the honey bees and bumble bees—as well as the stingless honey bees and orchid bees. These four groups are collectively called the corbiculate bees, but total less than 800 of the almost 6,000 species in this, the largest bee family.

Apidae are generally recognized by their long-tongued bee mouthpart morphology: a labial palpus with two long basal palpomeres and two short apical ones that are usually at right angles to the first two (see page 21). They also have a labrum that is narrowed somewhat basally, separating them from the other long-tongued bee family, Megachilidae, whose members have a labrum that is widest at the base. A few apids have reverted to the short-tongued bee labial palp structure.

The classification of the Apidae has undergone a lot of change since American melittologist Charles Michener's 2007 treatment of the bees of the world, and here, five rather than Michener's three subfamilies are recognized.

Anthophorinae comprises seven genera of large hairy bees with wings that are papillate apically, large hairless areas basally, a longer posterior margin to the first submarginal cell than that of the second, and mostly simple scopal hairs. Except for a couple of northern hemisphere species that nest in hollow cavities, these are ground-nesting bees. The subfamily includes almost 750 species found throughout the bee-inhabited parts of the world, although none of the genera are so ubiquitous.

Nomadinae can be identified as all cuckoo Apidae that lack the defining characteristics of Euglossini (e.g., a group of bristles replacing the jugal lobe, and found only in the western hemisphere tropics) or Xylocopinae that also do not collect pollen (a few eastern hemisphere tropical species). The 1,600-plus species are found on all continents except

Antarctica and collectively attack bees from almost all other major taxonomic groups. Approximately half of the species are found in the single genus *Nomada*. In superficial appearance, they are often very wasp-like bees, as in the image above.

Apinae contains the corbiculate bees and two genera of mostly oil-collecting species. The latter, *Centris* and *Epicharis*, share the wing characteristics of the Anthophorinae except that the posterior margin of the first submarginal cell is shorter than that of the second and the scopal hairs are largely plumose. All corbiculate bees (except for the kleptoparasitic and socially parasitic taxa) have a tibial corbicula in females: this means the metatibia is flat or somewhat concave on the anterior surface, where it is also largely hairless but with a fringe of hairs

on both the dorsal and ventral margins that serve to hold pollen in place.

Eucerinae was formally recognized as a subfamily only in 2018. Aside from two rare Argentinean exceptions that have a partial corbicula, they are non-kleptoparasitic, non-corbiculate Apidae without papillate apical areas to the wings and with a pygidial plate.

Xylocopinae includes the carpenter bees, along with two oil-collecting genera that seem to belong here. There are more than 1,000 species in this subfamily, distributed throughout the world. Most can be told from other apids by the combination of the presence of a scopa (albeit reduced in cuckoo and social parasitic forms), the absence of a pygidial plate or one that is reduced to a small spine hidden by the prepygidial fimbria, and a clypeus that is usually parallel-sided above and abruptly widened below. Xylocopinae nest within plant material, although one rare Middle Eastern genus and a few species of *Xylocopa* nest in the ground. Some species form small societies and a few are cuckoos or social parasites.

ABOVE | Large carpenter bees, like this *Xylocopa violacea*, have intimate relationships with mites, a large number of which can be seen near the base of this female's metasoma.

ABOVE LEFT | *Nomada* are among the most wasp-like of bees. This female *Nomada goodeniana* is at a host nest entrance.

AMEGILLA

Most species in this genus have pale markings on the face, mesosomal hairs that appear grayish brown, and white or blue apical hair bands to the terga. But there are exceptions, with more unusual color patterns—including species covered in bottle-green metallic hairs, others with largely golden pubescence, others with an entirely metallic blue metasoma, and still others with a couple of large areas covered in orange or white hairs. There are more than 220 species, giving plenty of scope for color variation.

One *Amegilla* species, Dawson's Mining Bee (*A. dawsoni*), is a very large species in which some males are much larger than the females. Males fight quite viciously over access to newly emerged females, often forming "mating balls" on the ground as a virgin female digs up to the surface from her subterranean natal home. Females can become seriously damaged because of the excessive attentions of overly amorous males. Smaller males are also produced, and these fly around flower patches frequented by foraging females, patrolling the edges of nest aggregations or arriving at the nest site earlier in the day than the large males. They find mates using a less combative approach.

GENUS
Amegilla

DISTRIBUTION
Throughout the eastern hemisphere except for the colder regions of northern Europe and northern Russia

HABITAT
Most commonly found in semiarid regions; during droughts they are sometimes active while most other bees remain quiescent

CHARACTERISTICS
- Scopa present on hind legs of female but not forming a corbicula
- Female with large pygidial plate
- First submarginal cell larger than second
- First recurrent vein arising near middle of second submarginal cell
- Third submarginal cell quadrate, anterior and posterior margins similar in length, as are the basal and apical margins
- Arolia absent

ABOVE | A mating pair of *Amegilla dawsoni*. The males of this species are combative in their efforts to find a mate.

ABOVE LEFT | Most *Amegilla* species have a banded metasoma with orange to gray hairs on the mesosoma.

RIGHT | A female *Amegilla dawsoni* at her nest entrance, checking that the surroundings seem safe enough for her to leave.

ANTHOPHORA

igger bees or flower bees are the common names sometimes applied to species in this genus. However, two of the 400-plus *Anthophora* species nest in cavities, so not all of them dig—further evidence of the difficulty of coming up with sensible common names. *Anthophora* are quite large bees, often with metasomal bands or other notable color patterns; a few species are mimics of bumble bees.

Females of at least some *Anthophora* species (perhaps all Anthophorinae) secrete fatty substances onto the walls of their brood cells and add some to the pollen ball. These chemicals come from the Dufour's glands, which are responsible for the shiny brood-cell lining in most ground-nesting bees and the cellophane lining in Colletidae nests. These glands are particularly enormous in *Anthophora* as a result of their nutrition-producing function. The brood cells of *Anthophora* smell "cheesy" because of these compounds. The larvae first eat the pollen ball and then consume the brood-cell lining. Thus, a substantial amount of the food the larvae require is provided directly from their mother's body rather than just her foraging activities.

Males of some *Anthophora* species have brushes or pads of hairs on the middle tarsi, which are passed over the female's head during mating, perhaps permitting the female to assess the quality of her suitor.

GENUS
Anthophora

DISTRIBUTION
Throughout the world except Australasia, Madagascar, and much of the Amazon and Congo Basins

HABITAT
Very diverse; semiarid areas to sparse woodland to rain forest

CHARACTERISTICS
- Scopa present on hind legs of female and not forming a corbicula
- Female with large pygidial plate
- First submarginal cell larger than second
- First recurrent vein arising near middle of second submarginal cell
- Third submarginal cell quadrate, anterior and posterior margins similar in length, as are the basal and apical margins
- Arolia present

BRACHYMELECTA

OPPOSITE | A male *Brachymelecta californica* forages on a flower. These bees need nectar to fuel their searches for host nests, but also need to feed on some pollen for protein to develop their eggs.

BELOW | A tray of specimens including the same "Nev." label as the specimen that confused melittologists for decades.

For 80 years a single bee sat in a collection identified as the only specimen ever found of its genus. It was a mystery because the only information available about where, when, and who collected it (standard information placed on labels) was a piece of paper that said nothing but "Nev." Unbeknownst to all melittologists, the same museum had a whole tray of the same species with the same imprecise locality information. But the larger sample had been identified as a different genus. Expert sleuthing by Canadian entomologist Thomas Onuferko resulted in the recognition in 2021 that the single specimen and those in the separate tray were in fact the same species. The single oddball was merely a partial albino that had two submarginal cells, whereas most, but not all, of those in the tray had three. As a result of the rules of nomenclature, the earlier generic name—*Brachymelecta*—takes precedence, even though it previously applied to only one specimen. The name that has been sunk, *Xeromelecta*, included six species and was known from perhaps thousands of specimens.

GENUS
Brachymelecta

DISTRIBUTION
Western North America, from southern Canada to southern Mexico and the Greater Antilles, with one individual seen in Virginia, USA

HABITAT
Diverse, from low altitude to over 10,000 ft. (3,000 m), in localities where the host species are common

CHARACTERISTICS
- Cuckoo Apidae
- Apex of midtibial spur lacking teeth
- Flabellum not divided
- Male scape at least twice as long as wide
- Marginal cell longer than stigma but not, or only slightly, extending beyond submarginal cells
- Inner ramus of tarsal claw of middle and hind legs short and broad, clearly different in shape to outer ramus

THYREUS

This is a genus of relatively large, handsome cuckoo bees, often with large patches of white to brilliant blue and occasionally purple appressed hairs on the mesosoma and especially the metasoma. These bright colors give them their common name of neon cuckoo bees. Identification keys to the species are difficult to use because they require distinguishing numerous shades of blue and careful assessment of the shape and size of the hair patches. Unfortunately, bee hairs get paler and sparser as the insect ages. There are more than 100 described species, from throughout Africa, Asia, and Australia, and much of Europe except the colder areas. They attack nests of the bee genus *Amegilla* (page 178) and, as with other Melectini, break open the completed cells of their ground-nesting hosts and lay their eggs inside before resealing them with moist soil.

BELOW | Some people call *Thyreus* neon cuckoo bees because of the bright blue hairs that often adorn their bodies as shown in this image. However, some are largely covered in snow-white pubescence.

GENUS
Thyreus

DISTRIBUTION
Throughout Africa, Asia (including many tropical islands), southern Europe, and mainland Australia

HABITAT
Diverse, arid scrub to wet forests where host species are common

CHARACTERISTICS
- Cuckoo Apidae
- Apex of midtibial spur lacking teeth
- Flabellum not divided
- Scutellum developed as a flat plate that extends posteriorly, its posterior margin concave
- Marginal cell rounded, not, or scarcely, extending beyond submarginal cells

CTENIOSCHELUS

D evil's coachwhip seems an appropriate common name for at least the males of these spectacularly beautiful bees: their antennae extend beyond the apex of the metasoma, giving the bees the superficial appearance of a cerambycid beetle. Presumably this feature is used to detect female pheromones or to physically interact with conspecifics. Both sexes of both species in the genus are patterned with gray, white, and black hairs on a metallic background, with the metasoma in particular being mostly greenish to turquoise in one species and bronze in the other.

These rare bees are thought to attack nests of bees of the genus *Centris* (page 196), most of which are oil-collecting bees. Phylogenetic studies indicate that kleptoparasitic (cuckoo) bees have rarely switched to oil-collecting hosts (perhaps only twice), and have a reduced speciation rate after the switch compared to cuckoos that attack hosts that use nectar instead of oils.

RIGHT | The *Ctenioschelus goryi* male is arguably the most dramatic-looking bee. While many other bees have a striking coloration, the enormous antennae combined with the beautiful color pattern of this bee make it quite special.

GENUS
Ctenioschelus

DISTRIBUTION
Mexico to Uruguay

HABITAT
Dry scrub and dry forest habitats

CHARACTERISTICS
- Cuckoo Apidae
- Labrum broader than medial length
- Flabellum divided into basal and apical portions
- Apex of midtibial spur toothed
- Male with enormously elongate antennae
- Scutellum bituberculate

OSIRIS

Most kleptoparasitic bees have a thick "skin" with pits and ridges to deflect the attacks of any hosts they might encounter. This is not the case in most of the 32 species of *Osiris*, which have a smooth and rather thin exoskeleton. Despite the lack of defensive armature, *Osiris* have very long stings that are made more effective through two additional mechanisms. First, S6 is strongly curved, as if used to guide the sting. Second, the furcula (which in almost all other bees never reaches outside the apex of the metasoma) is often exserted, seemingly giving additional reach for their weapon. There is evidence that the sting is used to kill the host egg.

BELOW | Compared to most of the cuckoo bees illustrated in this book (as on pages 187–93), this *Osiris* is rather slender and has a smooth, thin exoskeleton.

GENUS
Osiris

DISTRIBUTION
From central Mexico south to Paraguay and northern Argentina

HABITAT
Wherever host species are abundant

CHARACTERISTICS
- Cuckoo Apidae
- Labrum at least as long as wide
- Front coxa triangular with a transverse basal carina
- Female S6 curved to form tubular guide for sting, lacking bristles or spines
- Mandible with two subapical teeth
- Pygidial plate of female not attaining apical margin; of male, only apical

ISEPEOLUS

The genera *Isepeolus* and *Melectoides*, which together make up the tribe Isepeolini, are cuckoo bees that share rather complex modifications to the apex of the metasoma that serve to aid their kleptoparasitic attacks on the brood cells of other bees. In *Isepeolus*, the female S6 has a medial spine that in some species bears lateral barbs, whereas in *Melectoides* there is an almost spoon-like apical process on S6 that is surrounded by large, almost peg-like bristles. The apical terga are also modified, with lateral extensions and membranous flaps.

Isepeolus contains 12 species and *Melectoides* 13. Most species attack nests of *Colletes* (page 128), and their first-instar larvae have the sickle-shaped mandible commonly found in cuckoo bee larvae that eliminate all competition for the food the host has provided.

BELOW | *Isepeolus luctuosus* has a checkered color pattern, not just on its metasoma, but also on its head and mesosoma.

GENUS
Isepeolus

DISTRIBUTION
South America, from Colombia to southern Patagonia, but largely absent from the Amazon Basin

HABITAT
From very arid areas to temperate forests, and from sea level to 10,000 ft. (3,000 m)

CHARACTERISTICS
- Cuckoo Apidae
- Wings papillate beyond wing veins, cells with sparse hairs at most
- Arolia large
- Pygidial plate absent in both sexes
- Female S6 with apical spine or broader process
- Jugal lobe at least a quarter as long as vannal lobe

PASITES

These are cuckoo bees that attack Nomiinae, perhaps only *Pseudapis* (page 92) and its relatives. The egg is bent into a U-shape and inserted into the brood-cell wall, its blunt anterior end aligning with the inner surface of the cell wall. Its placement is undoubtedly aided by the narrow, apically concave female S6. *Pasites* have fine hairs compared to some Nomadinae, and these may be in attractive silky patches against a black or largely red background integument color.

Pasites is one of the few genera of bees in which the males have the same 10 flagellomeres to the antenna as do the females. The 26 known species vary from very small to more than ³⁄₈ inches (1 centimeter) in length.

GENUS
Pasites

DISTRIBUTION
From the Iberian Peninsula to eastern Russia, Africa, and Madagascar. Not known from tropical Asia east of India

HABITAT
Semiarid areas with healthy host populations

CHARACTERISTICS
- Cuckoo Apidae
- Mandibles directed mesally, apices overlapping not at all or at a very obtuse angle
- Labrum at least as long as broad
- Female S5 curved upward around S6 and sting, but otherwise unmodified
- S6 lacking bristles
- Male antennae with 10 flagellomeres

HOLCOPASITES

These bees have a distinctive checkerboard pattern of appressed white hairs against a red to red-and-black background. *Holcopasites* are cuckoo bees that attack nests of various panurgine genera including *Calliopsis* (page 68), *Pseudopanurgus*, *Protandrena*, and perhaps others. One species of the genus is known to learn the location of potential host nests and return to them time and time again, a behavior referred to as trap-lining. One species was once thought to be rare, but the "cuckoo bee task force" responsible for looking for it found that it was quite common at the larger nest aggregations of its *Calliopsis* host, successfully attacking up to 30 percent of the host brood cells.

BELOW | *Holcopasites* species have the most obviously checkerboard color pattern of any bee, with multiple white patches of hair on each metasomal tergum and red and black markings.

GENUS
Holcopasites

DISTRIBUTION
North America, from southern Canada to southern Mexico; seemingly absent south and east of the Isthmus of Tehuantepec

HABITAT
Wherever host species are abundant, usually at host nesting aggregations in sandy soil

CHARACTERISTICS
- Cuckoo Apidae
- Mid-coxa shorter than distance from its upper margin to base of hindwing
- Labrum at least as long as wide
- Episternal groove not extending below scrobal groove
- Female S6 apically concave and hairy
- Male with 10 flagellomeres
- T2–T4 with lamellate graduli that have a deep transverse furrow behind them

KELITA

This is a genus of southern South American cuckoo bees that attack a variety of different panurgine genera as hosts. For reasons unknown, cuckoo bees are often more difficult to identify to species level than are their hosts, and the numbers of species we know of are likely underestimated as a result. This is the case with *Kelita*—while five species have been described, at least an additional dozen are known.

Cuckoo bees lay eggs inside the host's nest, and their tail end is often highly modified as a result. Species-level characteristics can often be found in the details of the pseudopygidial and adjacent areas. This is the case with *Kelita*, where the shape of the pseudopygidial area, and the distribution of hairs of different lengths or shapes within it, are helping to delineate new species.

BELOW | A small *Kelita* female forages on a large aster flower. Cuckoo bee females spend most of the day looking for host nests. They are more often found on flowers late in the afternoon after the nests of hosts are closed for the rest of the day and hence inaccessible.

GENUS
Kelita

DISTRIBUTION
Western and Patagonian Argentina, and throughout Chile

HABITAT
Arid and semiarid areas where hosts are common

CHARACTERISTICS
- Cuckoo Apidae
- Labrum broader than long
- Procoxa triangular
- Female S6 bifurcate
- Scape less than twice as long as wide
- Tergal and sternal graduli not lamellate

EPEOLUS

The 134 *Epeolus* species are cuckoo bees whose hosts are all species of the cellophane bee genus *Colletes* (page 128). To lay their eggs, these bees enter incomplete host brood cells and cut holes in the cellophane cell lining using the teeth on a pair of processes on S6. These teeth are modified hairs. The egg is laid through this hole, which is then sealed by the action of Dufour's gland secretions that dissolve the brood-cell lining immediately around the hole and then dry to seal

the hole. The end of the egg against the cell lining is flat, making its existence difficult to detect through touch. As with many cuckoo bee larvae, the early instar has large mandibles to kill the host egg or larva and any additional cuckoo bee eggs or larvae.

Almost all *Epeolus* species have complex patterns of pale hair patches on a dark background (although some have large areas of red or orange cuticle). These markings are made from appressed, short,

LEFT | *Epeolus* species have patches of hair on both the meso- and metasoma, and details of the shape, size, number, and to some extent color, of these patches helps identify the species.

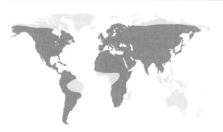

GENUS
Epeolus

DISTRIBUTION
Worldwide, but absent from Australia, islands in Southeast Asia and the Pacific, Madagascar, and Congo and Amazon basins

HABITAT
Wherever there are suitable hosts

CHARACTERISTICS
- Cuckoo Apidae
- Females have two spatulate, toothed lobes to S6
- Female pseudopygidial area is distinct, short, and usually lunate
- Male pygidial plate on one plane, apex U-shaped

broad hairs. The hair patterns are important for identification of the species. Some tropical western hemisphere species are entirely black, often with dark wings. The pseudopygidial area is short and usually semicircular; details of its form, and hair color and pattern, are also important for identification.

ABOVE | *Epeolus variegatus* is one of the more common species of the genus in Western Europe and it seems to attack a range of host species in the genus *Colletes*.

NOMADA

BELOW | Many bees sleep while holding on to a plant stem with their mandibles, and without using their legs, as in this *Nomada*.

With more than 750 described species, this is the largest genus of cuckoo bees in the world. Most of them are brightly colored black, with yellow and/or orange bands. Others have orange as the background color, with or without yellow markings. Still others are a darker red, sometimes with black and/or yellow markings.

GENUS
Nomada

DISTRIBUTION
Found on all continents except Antarctica, but scarce in tropical Africa, tropical Asia, and Australia, and absent from Madagascar, New Zealand, and Chile

HABITAT
Diverse; anywhere where the hosts are found, but especially common on sun-facing slopes with sparsely vegetated ground

CHARACTERISTICS
- Cuckoo Apidae
- Female S6 not curled into a tube or apically bifid, bearing some stiff bristles
- Labrum broader than medial length
- Apex of marginal cell pointed
- Wasp-like in coloration

It can take a beginner melittologists some time to realize that these very wasp-like insects are actually bees. As a general rule, any wasp-like insect that is flying low to the ground and often stopping to inspect the nest entrances of mining bees will be a species of *Nomada*.

Most *Nomada* attack mining bees of the genus *Andrena* (page 58), but a few attack bees of other families such as long-horned bees (*Eucera*; page 220), halictids, or even melittids.

Females find host nests through their sense of smell. It has been suggested that *Nomada* males transfer odors to their females during mating that make the female smell like a host *Andrena*. Some host cells may be attacked by multiple *Nomada* females, with the result that there can be more than one *Nomada* egg in the cell in addition to the host egg. The cuckoo bee larva that hatches first will kill conspecific eggs as well as that of the host before starting to feed on the pollen mass.

ABOVE RIGHT | Even cuckoo bees have to tank up on nectar to fuel their activities as in this *Nomada cornuta*.

RIGHT | As members of the largest genus of cuckoo bee, *Nomada* species are, unsurprisingly, variable in color. Compare this mostly red-and-black *N. femoralis* with the yellow-and-black *N. cornuta* above.

CENTRIS

BELOW | A female *Centris rhodopus*. Its Latin specific name means "red-legged," but the eyes also are red. Red eyes are not very common among the bees, although several other *Centris* species also have red eyes.

The genus *Centris* is made up of more than 245 species of generally large, attractive bees, most of which collect oil that is used in brood cell construction or as food for their offspring, or both. To obtain the oils they usually have complex patterns of elaborately structured hairs on the forebasitarsi and often on the mesobasitarsi, which are usually arranged into combs called elaiospathes. These are absent to weak in males and in females of species that do not collect oil, as well as those that collect oil from *Calceolaria* flowers (see page 215).

The nesting biology of *Centris* is variable, with some nesting in soil, others in wood, and some even in termite mounds. Brood cells are made with soil, moistened with nectar or floral oils. The oils are also used to line the brood cells, although the species that do not collect oils seem to use glandular secretions to line their brood cells.

RIGHT | Male *Centris pallida* digging up virgin females in Arizona, USA. Their camouflage coloration against the desert sand is enhanced by the dirt from their digging.

GENUS
Centris

DISTRIBUTION
Throughout the western hemisphere except Canada and most of the eastern USA, but present in Florida and the Caribbean, through Central America and south to Patagonia

HABITAT
Diverse, from deserts to forests

CHARACTERISTICS
- Hairy, long-tongued bees
- Stigma small, parallel-sided
- Labrum broader than long
- First submarginal cell small
- Scopal hairs mostly plumose
- Jugal lobe ½–¼ as long as vannal lobe
- Marginal cell shorter than distance between its apex and apex of wing

EUGLOSSA

GENUS
Euglossa

DISTRIBUTION
Tropical Central and South America;
one species has been introduced to
Florida

HABITAT
Mostly in damp tropical forests and
rain forests

CHARACTERISTICS
- Corbiculate bees
- Mostly bright metallic in color
- Jugal lobe replaced by a patch
  of bristles
- Labrum mostly pale in color
- Male metatibial slit short,
  not reaching tibial apex

This is one of five genera of orchid bee, including more than 130 species and so called because the males are attracted to orchids. They collect scents from the flowers using a mop of hairs on the front tarsi, and then transfer them for storage in special sponge-like tissues in their swollen hind tibiae. It is not just orchids the males visit for scents; they can be attracted to chemical baits and even DDT (not that they can find much of that these days, fortunately). They also collect volatiles from rotting fruit, fungi, and feces. The reason for this interest in olfactory diversity is that the males use the bouquet of odors to entice a female to mate; the more complex the odor, the more attractive the male.

Some species make clusters of brood cells inside preexisting cavities, while others make interesting nests that can be as simple as a dome over a cluster of brood cells or a more complex structure. The nest can be made from a variety of sources—including resins, which provide some additional waterproofing. The bees will even damage plants to increase the resin supply. One species has been observed stealing nesting materials from stingless bees. Nests are often occupied by a group of females that coexist in a communal society.

LEFT | This *Euglossa dilemma* female has a corbicula packed with pollen that has been moistened with nectar. This makes it less likely that she will lose any of this essential resource as she flies home.

RIGHT | Orchid bees such as this *Euglossa* usually have very long tongues, which increases the range of flowers from which they can obtain nectar.

EULAEMA

U nlike other genera in the tribe Euglossini, these large, bumble bee–like orchid bees do not have extensive metallic coloration. Instead, the 29 *Eulaema* species are black, usually with orange to yellow bands on at least some terga, although some are entirely black. Some have weak metallic colors on the metasoma only.

Nests are clusters of cells made from mud or feces mixed with resin or glandular secretions. The clusters are situated inside naturally occurring cavities. Females may occupy the nests in small communal groups.

As with other orchid bees, males collect fragrances as part of their courtship routines. Some *Eulaema* males bypass the process of obtaining fragrances from flowers by raiding the odor-storing structures of dead males!

LEFT | This handsome large *Eulaema cingulata* female has been collecting pollen, as can be seen on her metatibial corbicula, but she is now obtaining nectar from this flower.

RIGHT | The male near the bottom of this image has several orchid pollinia affixed to the back of her mesosoma.

GENUS
Eulaema

DISTRIBUTION
From Mexico to northern Argentina. Records from the extreme southern USA are likely males dispersing over long distances rather than persistent populations

HABITAT
Diverse, from semiarid areas to rain forests

CHARACTERISTICS
- Corbiculate bees
- A group of bristles replaces the jugal lobe
- Labrum dark
- Face not metallic, black, with or without pale markings
- Clypeus with a strong medial ridge

There are two genera of kleptoparasitic orchid bees, *Exaerete* and the large, shiny blue monotypic *Aglae*. Both genera have males that collect fragrances and store them in their hind tibiae, just like other orchid bees.

There are nine species of *Exaerete*; all are entirely metallic green to purple. They attack the nests of *Eulaema* (page 200) and *Eufriesea*, and their females open a sealed host brood cell, kill the host egg, and replace it with one of their own. They then close the site of their "break and entry" with similar materials to those used by the host. *Exaerete* females may leave the nest they are attacking upon the return of a host, but they will fight with conspecifics if a second kleptoparasite enters the nest.

As with all cuckoo bees, the distribution of *Exaerete* extends no further than that of their hosts.

LEFT | *Exaerete* are cuckoo orchid bees and usually just as brightly colored as their hosts.

RIGHT | This *Exaerete* male's tongue is longer than it needs to be to obtain nectar from this tropical flower in Brazil.

GENUS
Exaerete

DISTRIBUTION
From northern Mexico to Argentina

HABITAT
From semiarid areas to rain forests, wherever host species are abundant

CHARACTERISTICS
- Cuckoo orchid bees
- Bristles replace the jugal lobe
- Metafemur swollen
- Metatibia curved and swollen apically
- Scutellum with a swelling on either side

Despite the remarkable amount of attention that honey bees have received, there is still some controversy over the number of species. Some suggest as few as eight, others more than 10. One of the contentious species is the Himalayan mountain form *Apis laboriosa*, which might be called the "abominable honey bee." Like other species in the "giant" honey bee group, this species forms nests that hang down from structures, but unlike the others in its group, its colonies migrate, heading to higher-altitude meadows in summer and warmer, lower-altitude sites in winter.

In addition to the Western Honey Bee (*A. mellifera*), a second species has been domesticated: the Eastern Honey Bee (*A. cerana*). These two species, along with a couple of others, form the medium-sized group of honey bee species. Like their "giant" relatives, the small honey bees—such as the attractively patterned *A. florea*—do not make their nests inside a hive, but outside on a structure.

Many books have been written on this genus alone, and even more on *A. mellifera*, so there is no need (let alone space) to go into details about the remarkable biology of these bees here.

LEFT | *Apis florea* is one of the honey bee species that nests out in the open.

RIGHT | Workers of the giant honey bee *Apis dorsata* cover their brood combs with their bodies, several bees deep, forming an effective defense against natural enemies.

GENUS
Apis

DISTRIBUTION
Native to the warmer parts of Africa and Asia. The domesticated Western Honey Bee has been introduced throughout North and South America and Australasia for its honey, wax, and pollination services

HABITAT
Diverse in terms of aridity, temperature, and altitude; often the most abundant bees in intensive agriculture

CHARACTERISTICS
- Females with a corbicula
- Male with enormous eyes
- Both sexes with hairy eyes
- Hind tibial spurs absent
- Marginal cell extending relatively close to wing apex

These brightly colored bees are very familiar to people living within their geographic range. The diversity of color patterns among the 280-plus species is enormous, and there is sometimes a bewildering range of color forms within a species. More commonly, there is a confusing similarity of color pattern among species, even sometimes species from widely different parts of the bumble bee evolutionary tree. These are examples of Müllerian mimicry: where multiple species converge on the same color pattern so that only one individual among the entire suite needs to be tested by a predator before all bees of similar appearance are avoided. Some bees of other genera also mimic bumble bees for the same reason. However, most flies and beetles that have evolved to have a bumble bee–like coloration are relatively harmless to a predator and are examples of Batesian mimicry—where a harmless organism benefits by resembling one that will give a potential predator a nasty surprise.

Bumble bees are eusocial insects. Colonies are initiated in spring by a single female, which produces one or more broods of workers before the colony switches to producing males and the following year's nest foundresses (see page 33). However, some are social parasites that assume the role of a queen, whose workers then rear the usurper's offspring. The social parasite species do not have workers of their own.

LEFT | *Bombus terrestris* is common in Europe and has been domesticated for pollination, especially in greenhouses. Unfortunately, this has resulted in its being taken to other continents where it is not native, which has caused problems for some native bees.

GENUS
Bombus

DISTRIBUTION
Worldwide, except for sub-Saharan Africa and east of Wallaces' Line, and ranging as far north as northern Ellesmere Island; the Australian and New Zealand populations are introduced European species. More abundant in the northern hemisphere

HABITAT
Very diverse, from sea level to altitudes over 16,400 ft. (5,000 m); less abundant, but still present, in rain forests and deserts within their geographic range

CHARACTERISTICS

- Corbicula present in females except of the socially parasitic subgenus *Psithyrus*

- Marginal cell separated from wing apex by a distance similar to its length

- Jugal lobe absent

- Nonmetallic coloration

Few bees use anything other than pollen as the major source of protein for their brood, the exceptions being stingless bees. A few tropical western hemisphere species of the genus *Trigona* feed on carrion and a few more do the same on various detritus; these are called vulture bees and filth bees, respectively. In tropical Asia at least three species of meliponine, including two of *Lisotrigona* (which contains six species in total), obtain the proteins they need from the antibacterial enzymes in the tears of vertebrates. Swiss entomologist Hans Bänziger studied the attraction of these bees to his own eyes, finding that individual bees (which he marked with tiny dots of paint) returned up to 144 times to feed from his eyes in a single day.

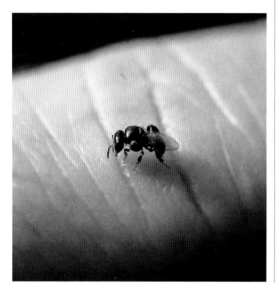

LEFT | Stingless bees often land on humans to imbibe sweat. They are called sweat bees in some parts of the world, whereas that common name is applied to bees in an entirely different family (halictids) in other parts of the world, which is more evidence for why the use of common names should be treated with caution. This is an image of *Lisotrigona carpenteri* taken in Vietnam.

GENUS
Lisotrigona

DISTRIBUTION
Tropical Asia, from India and Sri Lanka to the Malay Peninsula

HABITAT
Tropical dry scrub to moist forest

CHARACTERISTICS
- Corbiculate bees
- Sting nonfunctional, sting stylet indistinct
- Wing venation strongly reduced
- Body length less than ¼ in. (4 mm)
- Base of marginal cell acute
- Malar space shorter than diameter of flagellum
- Dorsoposterior angle of metatibia rounded

ABOVE | A cluster of *Lisotrigona* workers are seen here feeding on the lachrymal gland secretions of a Changeable Hawk-eagle (*Nisaetus cirrhatus*).

LEFT | A congregation of 20 *Lisotrigona furva* bees feasting on tears of an entomologist who photographed himself.

MELIPONA

The Spanish conquistadores burned almost all Mayan literature, but the few codices that remain have a remarkable amount of information about meliponiculture—the practice of keeping and rearing stingless bees. There are still people in the Yucatán Peninsula that practice the old methods of keeping these bees, and their knowledge has spread to many other parts of Mexico and Central and South America. In the northern parts of the genus's geographic range, *Melipona beecheii* is the most commonly managed species. Stingless bee honey is, in the author's

LEFT | Most *Melipona* species are brown and look quite a lot like stubby honey bees. This one, *Melipona quadrifasciata*, is quite different, however, perhaps looking more like a wasp than a bee.

RIGHT | A more standard *Melipona* can be seen in the *M. rufiventris* at the top right of this image, which shows a nest that has been partially opened to reveal the brood-cell comb within.

GENUS
Melipona

DISTRIBUTION
Mexico and much of the Caribbean to northern Argentina

HABITAT
Dry scrub to moist forest in the western hemisphere tropics

CHARACTERISTICS
- Corbiculate bees
- Sting nonfunctional but sting stylet right-angular to acute
- Wing venation strongly reduced; submarginal cross-veins weakly defined
- Wings scarcely attaining apex of metasoma
- Stigma in marginal cell weakly convex
- 8–14 hamuli

opinion, absolutely delicious, with a far more complex flavor than honey bee honey. It is sometimes used for medicinal purposes.

The 70 species in the genus *Melipona* are relatively large and some look remarkably like honey bees. However, like all other Meliponini they can be differentiated easily on the basis of the unusual reduced wing venation. Unlike other social corbiculate bees, including other meliponines, *Melipona* queens start off no larger than the workers (although they do end up a lot fatter once they start egg-laying on a large scale) and there is a genetic contribution to caste determination. Most species nest in cavities in wood, but at least one nests in the ground.

SCAPTOTRIGONA

There are 20 species in this genus of stingless honey bee, which have complex societies with queens and workers. New colonies are formed by fission of the old colony, with workers deciding on where the new nest should be. They travel backward and forward from the old nest to the new one, carrying nest construction materials and food with them. Eventually, the new queen is one of the individuals that flies to the new nest location and the independence of the new nest is gradually established. As she lays more and more eggs, the queen becomes physogastric—so heavy that she is incapable of flight.

LEFT | *Scaptotrigona postica* workers guarding their nest entrance.

GENUS
Scaptotrigona

DISTRIBUTION
Throughout South and Central America, but absent from Chile and Patagonian Argentina

HABITAT
Mostly forested areas in the western hemisphere tropics

CHARACTERISTICS
- Corbiculate bees
- Reduced wing venation
- Distinctly punctate
- Scutellum extends posteriorly as a thin shelf (in profile), the margin of which is rounded in dorsal view
- Shining V- or U-shaped depression on anterior margin of scutellum
- Anterior margin of pronotal lobe rounded, lacking carina

In the wild, *Scaptotrigona* nest in cavities in tree trunks and make horizontal brood combs, but the honey storage pots are more like bundles of grapes than the well-organized honeycomb of honey bees. Some species are kept in artificial hives for use in pollination or honey production.

ABOVE | A *Scaptotrigona bipunctata* forager approaches an iris flower that has already attracted another forager, perhaps her sister.

BELOW | Queen stingless bees end up as egg-laying machines incapable of flight, as in this physogastric individual.

EXOMALOPSIS

BELOW | *Exomalopsis* has a relatively flat, disk-like head, as can be seen in this male.

Bees in this genus of 88 species hold the record for the deepest nests: one nest studied by Brazilian entomologist Ronaldo Zucchi was 16 feet (5 meters) deep, yet the bees living in it were less than ³⁄₈ inches (1 centimeter) in length. The nest contained 884 females and 46 males, so it is probable that it had been built over multiple generations, with each generation comprising multiple females. *Exomalopsis* live in communal societies, most of them a lot smaller than the one excavated by Zucchi, with 20 individuals being on the large size. Females of one of the cuckoo genera that attack *Exomalopsis* have unusually flattened bodies, presumably so they can press themselves against the burrow wall and hence reduce their chances of detection by the many hosts passing by.

GENUS
Exomalopsis

DISTRIBUTION
From central USA to central Argentina, with one species in northern Chile

HABITAT
Diverse, from semiarid areas to grass and scrublands, and dry forests to rain forest

CHARACTERISTICS
- Long-tongued bees
- Second abscissa of vein M+Cu of hindwing more than twice as long as cu-v
- Clypeus not strongly protuberant
- Row of long, well-separated hairs along inner eye margins
- Scopal hairs strongly plumose
- Female basitibial plate large

This genus belongs to the Tapinotaspidini, a tribe of oil-collecting bees now considered to belong to the subfamily Eucerinae. There are 21 species in this genus. The oil is mixed with pollen and carried back to the nest in the scopa on the hind legs. Many species obtain their oil from flowers of the plant genus *Calceolaria*, which are commonly bright yellow, but sometimes blue or purple. The large, bucketlike lower petal is the one where oils are secreted.

In one study, over 80 percent of the visits to a *Calceolaria* species were performed by a single species of *Chalepogenus*, and the plants set no seed if pollinators were excluded.

BELOW | A female *Chalepogenus rozeni* has oil-moistened pollen in her hind leg scopa, having obtained both the oil and the pollen from this *Calceolaria* plant.

GENUS
Chalepogenus

DISTRIBUTION
Western South America from Ecuador to Patagonia and eastern South America from southern Brazil southwards

HABITAT
Semi-arid and scrub to moister areas

CHARACTERISTICS
- Dense plumose hairs beneath the long simple hairs on the metatibia and metabasitarsus
- Second abscissa of vein M of hindwing less than twice as long as cu-v
- Inner surface of front basitarsus convex; outer surface covered by a dense pad of finely branched hairs
- Mesoscutum with a few long hairs
- Middle tibial spur not notched near apex; inner hind tibial spur not curved at base
- Vertex behind ocelli not carinate

DIADASIA

OPPOSITE | *Diadasia ochracea* is an appropriate name for this ocher-colored bee. The species collects pollen from plants in the Malvaceae family, but is clearly not collecting pollen here.

BELOW | A female *Diadasia* at her nest entrance. The slightly raised rim around the entrance is the beginnings of a turret.

This genus of 40 described species of western hemisphere bees is distributed from southern Canada to Chile and Argentina, although it is largely absent from the Amazon Basin. It is one of the better-studied genera of bees from the perspective of the evolution of floral specialization, with many species foraging from mallows, which were probably the ancestral hosts. One lineage switched to cacti, another to asters, and one species each to *Clarkia* and a bindweed.

Their nests are made in the ground and are relatively shallow and near the surface, although they often include turrets at the entrance, which are sometimes downturned. This gives them their suggested common name of turret bees.

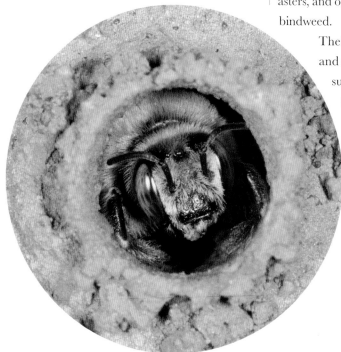

GENUS
Diadasia

DISTRIBUTION
Throughout the western hemisphere except the far north of Canada, much of eastern North America, the Caribbean, Patagonia, and the Amazon Basin

HABITAT
Diverse, from semiarid areas to deciduous forests, but seemingly largely absent from rain forest

CHARACTERISTICS
- Eucerinae that do not collect oils
- Paraglossa short
- Male posterior claws usually broad and bluntly rounded
- Male metapostnotum usually bare
- Gradulus of S2 strongly convex posteriorly, angulate to lobate
- Vertex convex in facial view

PTILOTHRIX

BELOW | *Ptilothrix* such as this eastern *Ptilothrix bombiformis* can walk on water, as shown here. They imbibe water, using it to soften the soil in which they nest.

Many years ago, the author was with some melittologists digging bee nests in the Arizona desert. All of a sudden one of us struck water—the shovel had broken a water pipe. Within minutes a large puddle formed and began attracting thirsty insects—primarily some wasps and bees of the genus *Ptilothrix*, all of which were landing on the water surface, legs outstretched to spread their weight as widely as possible so as to not break the surface tension of the water.

Ptilothrix is a genus of 16 species that make nests in the ground. Some species moisten the soil with water they collect, making it easier to excavate and construct brood cells underground and turrets at the entrance.

GENUS
Ptilothrix

DISTRIBUTION
The New World, from southern Canada to northern Patagonia; absent from Chile and around the equator

HABITAT
Diverse, from deserts to temperate woodlands; largely absent from rain forest

CHARACTERISTICS
- Eucerinae that do not collect floral oils
- Proboscis in repose not reaching further than front coxae
- Arolia absent
- Vertex convex in facial view
- Male antennae relatively short
- Stigma longer than prestigma

EUCERINODA

One species in this genus, which is the only member of its own subtribe, is the sister species to the other 800 or so species of Eucerini, from which it separated more than 50 million years ago. This species was thought to be extinct, having not been seen for 50 years, until the author found a population on the road between Santiago, Chile, and the main ski resort nearby. It has since been found more than 230 miles (750 kilometers) further north, but only a single specimen, and at an altitude above 10,000 feet (3,000 meters). The males have interesting secondary sexual modifications, and the females seem to forage mostly early in the morning, with a preference for Asteraceae.

BELOW | Unlike almost all other bees in the long-horned bee tribe Eucerini, *Eucerinoda gayi* males do not have unusually long antennae. However, they do have interestingly modified hind legs, including the basitarsus, as can be seen here.

GENUS
Eucerinoda

DISTRIBUTION
Central Chile, from near Santiago to 500 miles (750 km) north

HABITAT
Semiarid areas within its geographic range. Likely no longer present at lower-altitude locations—recent records are from above 6,500 ft. (2,000 m)

CHARACTERISTICS
- Jugal lobe of hindwing at least half as long as vannal lobe
- Paraglossa much shorter than first segment of labial palpus
- Male with pale markings on lower paraocular area and lacking metatibial spurs
- S2 gradulus simple, not biconcave

While most recent changes to bee taxonomy have involved splitting one genus into two or more, *Eucera* has undergone the opposite trend, with multiple previously recognized genera now subsumed within it (although these changes are not, as of 2022, universally accepted). As a result, the genus has grown from around 220 species to more than 380. Among the newly assigned species are the squash bees, previously the genera, and now subgenera, *Peponapis* and *Xenoglossa*. The Hoary Squash Bee (*Eucera pruinosa*) is one of the best-known wild bees in North America, where it is an important pollinator of Zucchini and other squashes, even without management. This species tumbles in and out of flowers early in the morning; the males sometimes sleep in the wilting flowers, which last only a day, at the end of hours of searching for females. The spread of this species in the USA with the northward and eastward expansion of squash cultivation from its origins in the southwest has been documented with genetic data.

More broadly, the genus *Eucera* is the original long-horned bee genus, so called because of the enormously elongate antennae of the males.

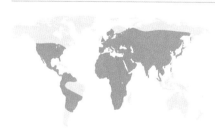

GENUS
Eucera

DISTRIBUTION
Worldwide except for tropical islands in east Asia, Australasia, the Amazon Basin, Patagonia, and the far north

HABITAT
Diverse, from semiarid areas to woodland, and including agricultural fields

CHARACTERISTICS
- Long-horned Apidae
- Paraglossa as long as first two labial palpomeres combined
- Female S2 gradulus strongly biconvex
- Male T7 lacking lateral teeth

ABOVE | This *Eucera* male demonstrates the reason for the common name applied to this genus and its relatives—the long-horned bees.

OPPOSITE | A female *Eucera nigrescens* approaches a thistle.

While the males of almost all other bee genera have antennae that are longer than those of the females— at least partly to make their chemosensory detection of female sex pheromones more efficient—it is in the long-horned bees that this is perhaps most evident (although see also *Ctenioschelus*, page 185).

THYGATER

THYGATER

BELOW | It is relatively uncommon for bees to have a labrum that is paler than the clypeus, but this feature can easily be seen in this *Thygater*.

Males in this genus have antennae that are long even for long-horned bees—when folded back over the body, they reach toward the end of the metasoma. Another unusual feature of the males is that while the clypeus is usually black, the labrum is usually pale, often white.

GENUS
Thygater

DISTRIBUTION
From Mexico south to Argentina, but absent from southern Argentina and Chile

HABITAT
Diverse, from scrubland to rain forest (although less abundant in the latter)

CHARACTERISTICS
- Long-horned bees
- Paraglossa as long as first two labial palpomeres combined
- Female S2 gradulus weakly biconvex
- Female scape only twice as long as wide
- Male T7 apically concave, usually with two teeth or lobes

These are relatively large, ground-nesting bees that are common in the western hemisphere south of the US–Mexico border. Look out for the first record of this genus from the USA as the climate warms.

Most bees have six maxillary palpomeres, but this number has been reduced in many lineages, and in the long-horned bees the number varies considerably, from the ancestral six down to two. *Thygater* has three or four maxillary palpomeres.

Some of the genera closely related to *Thygater* are extremely rare.

CTENOPLECTRA

Bees nesting in hard wood are often distinctly cylindrical, efficiently occupying a tubular nest. This is clearly the case for species of *Ctenoplectra*, which have been observed nesting in old beetle burrows in hard wood, and also inside old brood cells of megachilids. Seventeen of the 19 species collect oils for their offspring, the other two being cuckoo bees that attack the nests of their congeners. The bees collect oil from cucurbits using straggly, mop-like groups of hairs on the metasomal sterna. The name *Ctenoplectra* refers to the triangular metatibial spurs that squeeze the oil into the brood cells when the bees return to their nest. Recent DNA data suggest that these bees are related to carpenter bees (page 228) rather than belonging to the Apinae where they were classified until recently, and some experts probably think they should have stayed there.

GENUS
Ctenoplectra

DISTRIBUTION
Sub-Saharan Africa, east Asia as far as southeast Russia, and northeastern Australia

HABITAT
Warm, humid areas with abundant *Mormodica* or *Thladiantha* floral resources

CHARACTERISTICS
- Cylindrical body form
- Stipital cavity and comb present
- Galeal comb absent
- Short labial palpi and glossa
- Medially divided transverse patches of straggly hairs on at least S3–S5

TETRAPEDIA

This is another genus of oil-collecting bees that DNA data suggest should be allied with carpenter bees (page 228). Oil-collecting structures in *Tetrapedia* include a dense probasitarsal comb, a robustly pectinate hind tibial spur, and a scopa with fine long branches to most of the hairs but longer, simple "guard" hairs to assist with oil transport. Unusually, males also possess some of these structures and collect oil in an identical manner to that employed by females. It has been suggested that the males use chemical deception, smelling like the floral resources the females are attracted to. *Tetrapedia* nest in previously formed tunnels in wood, and frequent nest usurpation and abandonment have been documented, although females seem not to disperse very far. Twenty-five species are known.

LEFT | *Ctenoplectra* are unusually cylindrical bees.

RIGHT | The fluffy hind legs of this male *Tetrapedia*, which they use to transport floral oils, are quite similar to the scopa found in females.

GENUS
Tetrapedia

DISTRIBUTION
From Mexico to southern Brazil and northern Argentina; absent from Chile

HABITAT
Rain forest to scrub forest where abundant oil-producing plants can be found

CHARACTERISTICS
- Oil-collecting Apidae
- Forebasitarsus with a comb
- Marginal cell bent away from anterior wing margin for much of its length
- Jugal lobe very small, less than ¼ as long as vannal lobe
- Pedicel at least 1.3 times as long as wide and scape short
- Arolia absent
- Metepisternal pits close together but not united as one large pit

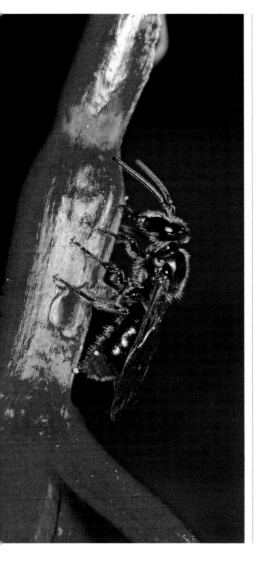

There are three species in this southern South American genus, which is the only member of its tribe. One species, *Manuelia gayi*, is all blue, while the other two are black—although *M. postica* has a red tail. The third species, *M. gayatina*, is the smallest of the three. Males of all species have pale facial markings.

These bees are solitary, although nests containing multiple females are sometimes found. A type of native bamboo is a commonly used host plant. There seems to be strong selection for larger mandibles in females that nest in this plant, but not in those that use the softer stems of introduced wild blackberries.

The genus is only known from Chile, south of La Serena, and Patagonian Argentina (see map). This region harbours a large number of endemic taxa in a large number of different animal and plant groups.

LEFT | A *Manuelia postica* male showing off the red tip to his metasoma.

RIGHT | *Manuelia* are shiny, relatively bald bees that transport pollen back to their nest with the hairiest part of their body, their metatibiae.

GENUS
Manuelia

DISTRIBUTION
Chile, from the southern margin of the Atacama Desert to Patagonia, and Patagonian Argentina

HABITAT
Semiarid to somewhat moist areas, generally not at high altitude

CHARACTERISTICS
- Three submarginal cells
- Stigma present
- Female with spine-like pygidial plate
- Jugal lobe of hindwing less than a fifth as long as vannal lobe
- Female with scopa on metatibia

XYLOCOPA

OPPOSITE | *Xylocopa* are often remarkably sexually dimorphic. This male *X. aestuans* is uniformly pale in color, whereas the female is black with golden-yellow hairs on the mesosomal dorsum.

BELOW | When bees have blue on their bodies, whether on hairs or the integument, the color is metallic, resulting from reflections rather than blue pigment. However, in a small group of *Xylocopa* species, such as this *X. caerulea*, the blue hairs do result from pigment.

Known as large carpenter bees, these impressive insects come in a wide range of colors and patterns. The common species in eastern North America has a yellow-haired thorax but is otherwise dark, the common European species is entirely black with brilliant violet wings, and a group of tropical Asian species has blue hairs—uniquely among the bees, this is caused by a blue pigment rather than a metallic effect. Males of many species are entirely tan in color appearing completely different from the females.

As their name suggests, carpenter bees make their nest in wood or woody substrates such as old flowering stalks of agave or sotol (see page 26). For some species, fights at nest entrances are common—the resource inside being a tunnel that would have taken an extremely long time to excavate by chewing. Some females wait in their natal nest for a year or two, "hoping" to inherit it on their mother's demise.

The genus includes 376 species, some of which are important pollinators of passion fruit. In some *Xylocopa* species the first metasomal tergum is invaginated as an enormous cavity that houses mites; these are beneficial to the bees as they consume fungal mycelia that develop in the brood cell.

GENUS
Xylocopa

DISTRIBUTION
Throughout the world, including many tropical islands, but absent from New Zealand, some small islands, and colder regions of the far north

HABITAT
Diverse, but absent from areas with insufficiently numerous large pieces of wood or robust stems for nest-building

CHARACTERISTICS
- Large bees with well-developed wing venation
- Marginal cell long and narrow
- Stigma absent
- F1 as long as F2 and F3 combined
- Arolia absent

EXONEURELLA

In most bees each offspring is individually housed in its own brood cell. However, this is not the case for bees of the tribe Allodapini, such as *Exoneurella*. These bees place their brood at the end of a burrow, usually in a twig or other hollow in plant material, and they are all mixed together, from egg through larva to developing pupa. The larvae have to move around to obtain the food that is brought to them indirectly, often after having been stored around the inside of the nest burrow closer to the entrance. As a result, the structure of the larvae is unusually varied, with false leg-like bumps and patches of hair, which can get particularly long and unkempt around the head.

OPPOSITE | Like many bees in the related genus *Braunsapis*, this *Exoneurella setosa* is almost entirely black but with a white marking on the clypeus.

BELOW | *Exoneurella eremophila* is the most boldly patterned species in its genus.

GENUS
Exoneurella

DISTRIBUTION
Throughout much of Australia, but absent from mountains and humid tropics

HABITAT
Arid and semiarid areas where suitable nesting materials can be found

CHARACTERISTICS
- Two submarginal cells
- Hairs relatively sparse
- Second recurrent vein absent
- Clypeus somewhat parallel-sided above, broadened below, constricted at level of anterior tentorial pits to give somewhat hourglass shape
- Mesosoma and metasoma largely black (occasionally with pale metasomal bands)
- Basitibial plate absent

There are four species of *Exoneurella*, one of which, *E. tridentata*, is so named because the apex of the metasoma of the largest females bears three points. The species varies considerably in size because it is eusocial, with workers that are half as large as their queen. It nests in hollowed-out dead twigs of Western Myall (*Acacia papyrocarpa*) and Boonaree (*Alectryon oleifolius*) shrubs, and is the only known truly eusocial xylocopine. Other species of the genus have small casteless societies, in which multiple females work in the same burrow without a strict hierarchy, but with some females waiting to become egg-layers.

There are 370 described species of small carpenter bee. They come in a wide range of colors—mostly dull metallic blue or green, but some are spectacularly bright purple, others are largely black, and many have complex patterns of yellow markings on the head, mesosoma, and metasoma. One species has such convoluted yellow markings on its body that it has been given the name *Ceratina hieroglyphica*. The genus is found on all continents except Antarctica, but diversity decreases substantially in the far north and only one species is known from Australia.

Ceratina species nest in narrow holes in wood or in pithy stems—often those that have been excavated by some other insect, although they are also capable of digging their own nests into softer dry plant material. After the female has completed her burrow, she will collect pollen and nectar, make a pollen ball, lay an egg, and then construct a partition from the material in which

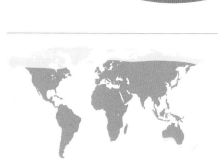

GENUS
Ceratina

DISTRIBUTION
Almost worldwide except New Zealand and the far north of both eastern and western hemispheres

HABITAT
Diverse

CHARACTERISTICS
- Stigma present
- Female with scopa on metatibia
- Jugal lobe ⅓ as long as vannal lobe
- Pygidial plate and fimbria absent in females
- Three submarginal cells
- Sides of upper part of clypeus subparallel

she is nesting. The resulting brood-cell divider looks like a disk of fine sawdust.

Some *Ceratina* species habitually share nests with two or more females in the same tunnel. In such cases, one female often does most of the foraging, the other staying behind to guard the nest. The forager sometimes eats the eggs laid by the guarding bee (which is usually larger) and replaces them with one of her own. A few species of *Ceratina* do not produce males, the females reproducing parthenogenetically.

Many of the parts of a bee described here are shown in the diagrams on pages 14–21.

Abdomen: The third and most posterior body part of an insect, considerably and misleadingly modified in bees (see page 14).

Abscissa: A section of a wing vein between two nodes.

Anterior tentorial pit: A small pit that denotes the attachment of the tentorium to the anterior surface of the head.

Apodeme: An extension of the external skeleton that projects into the body and serves as a site for muscle attachment.

Appressed: Hairs are said to be appressed if they lie flat against the body.

Arolium (p = arolia): A lobe between the claws of the apical tarsomere.

Basitarsus (p = basitarsi): The basal tarsomere.

Basitibial plate: A usually small, usually triangular area at the base of the *metatibia* on the dorsal surface.

Carina (p = carinae): A narrow, raised line.

Claws: The apical tarsomere.

Clypeus: The facial sclerite between the epistomal sulcus and *labrum*.

Communal: A form of social organization in which female bees live together in a nest without any division of labor.

Compound eye: The large eyes on the side of the bee head, made up of many ommatidia (facets).

Corbicula (p = corbiculae): A bare area surrounded by hairs that serves to make a pollen basket. Usually considered as the anterior surface of the *metatibia* in corbiculate bees, but bare areas surrounded by hairs can be found on other body parts in other bees.

Coxa (p = coxae): The basal segment of the leg, articulating with the thorax and the *trochanter*.

Crepuscular: Active under low light conditions, such as around sunset or dawn.

Cross-veins: Veins that run between longitudinal veins.

cu-v: A *cross-vein* between veins M+Cu and V in both forewings and hindwings.

Diapause: A period of suspended development that permits survival of unfavorable conditions with specific initiating and terminating conditions.

Dufour's gland: A gland in the *metasoma* that secretes chemicals usually associated with brood cell linings.

Episternal groove: A groove on the mesopleuron extending vertically from a pit

beneath the base of the forewing.

Epistomal lobe: A convexity in the epistomal sulcus that extends onto the *clypeus*.

Eusocial: Cooperative brood care with a reproductive division of labor between generations; in bees, this means the workers are typically the daughters of the queen.

Exoskeleton: The external skeleton typical of insects and other arthropods.

Facial fovea (p – facial foveae): A concave area on the upper face marked by differences in surface sculpture and/or setation.

Falcate: Hooked.

Femur: The third portion of a leg, between the trochanter and tibia.

Flabellum: A usually disc-shaped structure at the apex of the glossa in some bees.

Flagellum (p = flagella): The antenna beyond the *pedicel*: in bees, usually comprising 10 flagellomeres in females and 11 in males.

Floccus: A tuft of long, curved hairs arising from the hind *trochanter* that closes the femoral *corbicula* at its base.

Furcula: A small, usually Y-shaped structure at the base of the sting shaft.

Galea (p = galeae): The apical part of the *maxilla*, beyond the *stipes* and *maxillary palp*.

Galeal comb: A row of bristles arising from a convexity on the *galea*.

Gena (p = genae): The area behind the *compound eye* in lateral view.

Glossa: A lobe from the labium that forms the tongue proper.

Glossal rod: A longitudinal thickening inside the *glossa* of most long-tongued bees.

Gonobase: A usually somewhat ring-shaped structure at the base of the genital capsule.

Gonocoxa (p = gonocoxae): Paired structures between the *gonobase* and *gonostyli* usually making up the largest part of the genital capsule.

Gonostylus (p = gonostyli): A sclerotized lobe usually at the apex of the gonocoxa.

Gradulus (p = graduli): A posteriorly oriented *carina* or lamella on a metasomal *tergum* or *sternum*.

Hamuli: A series of hooks on the anterior margin of the hindwing that interlock with a recurved posterior portion of the forewing so that the two wings on each side beat in unison in flight.

Hypostomal area: The area lateral to the oral fossa, behind the mandibular articulation.

Integument: The body's external covering.

Jugal lobe: The convex basal portion of

the posterior margin of the hindwing that is delimited distally by a notch in the wing margin.

Keirotrichia: Short, usually blunt or bifid hairs on the posterior surface of the hind *tibia*. The keirotrichiate areas of each hind tibia serve to clean the wing surfaces.

Kleptoparasite (also spelled cleptoparasite): An insect that steals food from another; in bees this refers to the habit of laying eggs on pollen masses collected by a different species.

Labial palp (p = labial palpi): A seemingly segmented structure, part of the *labium* and lateral to the *paraglossa*, usually comprised of four *palpomeres*.

Labium: The most posterior portion of the mouthparts. It includes the *prementum*, *labial palpi*, *lorum* and *mentum*.

Labrum: A sclerite that is attached to the apical margin of the *clypeus*.

Lacinia (p = laciniae): A small, often hairy flap usually located anteriorly between the *stipes* and *galea*.

Lamella: A thin plate of *exoskeleton*, somewhat blade-like.

Lorum: A V- or Y-shaped sclerite at the base of the *labium*.

Mandible: The insect "jaw," articulating with the head capsule.

Marginal cell: The cell beyond the *stigma* along the anterior margin of the forewing.

Marginal zone: The portion of a *tergum* or *sternum* toward the apex that is usually of thinner cuticle than most of the rest of the structure. Also called the apical impressed area.

Maxilla: The part of the *proboscis* that is behind the *mandible* but in front of the *labium*.

Maxillary palp (p = maxillary palpi): A seemingly segmented structure arising between the *stipes* and *galea* on the posterior surface of the *maxilla*, usually comprising six *palpomeres*.

M+Cu: The longitudinal vein between the radial and vannal (anal) veins on both the forewing and the hindwing.

Mentum: That part of the *labium* between the *lorum* and *prementum*.

Mesosoma: The middle tagma of the bee body, between the head and *metasoma*, bearing the wings and legs.

Metafemur: The *femur* of the hind leg.

Metanotum: The dorsal sclerite of the third thoracic segment, between the *scutellum* and *metapostnotum*.

Metapostnotum: The sclerite between the *metanotum* and *propodeum*.

Metasoma: The last tagma of the bee body, consisting of the *abdomen* minus the first segment.

Metatibia: The *tibia* of the hind leg.

Metepisternal pits: Relatively large pits on the metapleuron.

Mimicry: Resemblance to a different species.

Notaulus (p = notauli): A short submedian line, usually impressed, sometimes made up of crowded punctures, that arises on the anterior margin of the *scutum*.

Ocellus (p = ocelli): Three isolated facets on the dorsal surface of the head between the *compound eyes*.

Oligolectic: Foraging for pollen on one or only a few closely related plant species. Also called specialist feeders.

Omaulus: A vertically oriented *carina* on the mesopleuron that extends ventrally from near the *pronotal lobe*.

Palpomere: The *labial* and *maxillary palpi* are divided into seemingly segmented structures, more properly termed palpomeres than segments.

Papilla (p = papillae): A nipple-shaped protrusion from a surface.

Paraglossa (p = paraglossae): A structure between the *glossa* and the *labial palp*.

Paraocular area: The facial area medial to the inner margin of the *compound eye*.

Parapsidal line: An impressed line between the *notauli* and *tegula*.

Parthenogenesis: Virgin birth.

Pectinate: Having long, narrow, close-set teeth.

Pedicel: The second segment of the antenna between the *scape* and *flagellum*.

Penis valve: A pair of usually long, narrow sclerites that form the apicomedial portion of the genital capsule.

Physogastric: The swollen *metasoma* found in queens of some *eusocial* insects.

Plumose: Branched, somewhat feather-like.

Polylectic: Foraging for pollen on a range of distantly related plant species. Also called generalists.

Premental fovea: The depressed area of the *prementum* of some bees margined by a *carina* or ridge, and usually filled with spicules or setae.

Prementum: The major part of the *labium*, located between the *mentum* and *labial palpi*.

Prepygidial fimbria: A transverse patch of dense hairs toward the apex of T5 in females.

Prestigma: A thickened portion of vein R basal to the *stigma*.

Proboscis: The feeding tube formed from the *maxillae* and *labium* of the mouthparts.

Pronotal lobe: A rounded, swollen posterolateral portion of the *pronotum* that lies over the first thoracic *spiracle*.

Pronotum: The dorsal sclerite of the prothoracic segment.

Propodeum: The first abdominal *tergum* that is fused to the posterior margin of the thorax and appears to be part of the thorax.

Pseudopygidial area: The apicomedial portion of T5 in females that is modified in Halictini, some kleptoparasitic Apidae, and a few other bees.

Pygidial plate: The raised apicomedial portion of T6 of many female bees and T7 in some males.

Recurrent vein: Usually a pair of veins, each of which arises on the posterior margin of a *submarginal cell* extending posteriorly.

Scape: The basal segment of the antenna.

Scopa (p = scopae): Hairs used to transport pollen.

Scrobal groove: An approximately horizontal groove on the mesopleuron posterior to, and usually joining, the *episternal groove*.

Scutellum: A sclerite between the *scutum* and *metanotum*, a part of the second thoracic notum. Also called the mesoscutellum.

Scutum: The major part of the dorsal surface of the *mesosoma*. Also called the mesoscutum.

Semisocial: Cooperative brood care with a reproductive division of labor among individuals of the same generation.

Serrate: Toothed or undulate, the teeth broader than they are long.

Social parasite: A species that invades the nest of a species with queens and workers, and supplants some or all of the egg-laying of the queen with her own. The offspring are raised by the workers of the host species.

Solitary: With only a single female living in each nest.

Spiracle: The small, round exterior openings to an insect's gas-exchange system.

Sternum (p = sterna): The ventral sclerite of a segment.

Stigma: A sclerotized area on the leading edge of the forewing, its posterior margin

forming an anterior margin of the first *submarginal cell* and anterobasal margin of the *marginal cell*.

Sting: A complex of structures associated with the sting shaft, which is the part of the sting through which the venom passes.

Stipes (p = stipites): Part of the *maxilla* basal to the *galea*.

Stipital cavity: A concavity in the posterodistal margin of the *stipes*.

Stipital comb: A row of bristles in the *stipital cavity*.

Subantennal sulcus (p = subantennal sulci): A depressed line that runs from the antennal socket to the epistomal sulcus.

Submarginal cells: From one to three wing cells that are margined anteriorly by the *prestigma*, *stigma*, and *marginal cell*.

Sulcus (p = sulci): A groove that is the external manifestation of an internal ridge.

Supraclypeal area: A triangular area above the *clypeus* and between the antennal sockets, demarcated laterally by the *subantennal sulci*.

Tarsus (p = tarsi): In bees, the apical portion of a leg that is divided into five tarsomeres, the *basitarsus* being the first.

Tegula (p = tegulae): A dorsally convex, usually oval sclerite that is at the base of the forewing.

Tergum (p = terga): The dorsal sclerite of a segment, synonymous with notum, although in melittology tergum is used for the *metasoma* and notum for the *mesosoma*.

Tibia (p = tibiae): The fourth segment of the leg, between the *femur* and *basitarsus*.

Tibial spurs: Spurs at the apex of a *tibia*; there is usually one midtibial spur and a pair of hind tibial spurs. The anterior tibial spur forms part of the antenna-cleaning apparatus.

Trochanter: The second segment of a leg, between the *coxa* and *femur*.

Tuberculate: Bearing tubercles, small knob-like protrusions.

Tumulus (p = tumuli): The pile of earth at the entrance of a nest in the ground that forms from the soil brought to the outside of the nest during excavation.

Vannal lobe: A convexity in the posterior margin of the hind wing toward the base, margined basally and apically by notches in the wing margin. The *jugal lobe* is basal to the vannal lobe.

Vertex: The dorsal margin of the head in frontal view.

BEE BIOLOGY AND NATURAL HISTORY

Danforth, B. N., R. L. Minckley, and J. L. Neff. 2019. *The Solitary Bees: Biology, Evolution, Conservation.* Princeton University Press, Princeton, NJ.

Michener, C. D. 1974. *The Social Behavior of the Bees: A Comparative Study.* Harvard University Press, Cambridge, MA.

Packer, L. 2010. *Keeping the Bees.* HarperCollins, Toronto.

TAXONOMIC LITERATURE

Bee taxonomy is constantly changing, and it is likely that almost no two melittologists will agree on exactly the same classification for all the bees. The classification followed in this book is based on personal decisions about often controversial views.

John Ascher and John Pickering's Discover Life bee website is a remarkable resource that was essential for estimating the number of bee species in different groups and for informing the creation of the distribution maps. Its citation is: Ascher, J. S. and J. Pickering. 2006–2022. Discover Life bee species guide and world checklist (Hymenoptera: Apoidea: *Anthophila*). [Draft 55, 17 Nov 2020].

Bee species guide: http://www.discoverlife.org/mp/20q?guide= Apoidea_species&flags=HAS. It also includes keys to the identification of genera and species, especially for eastern North America, as well as lists for all countries.

Bee species world checklist: http://www.discoverlife.org/mp/ 20q?act=x_checklist&guide=Apoidea_species

Images of almost all of the world's bee genera can be found on one of the author's York University web pages at https://www.yorku.ca/bugsrus/resources/galleries/bgow

Michener, C. D. 2007. *The Bees of the World,* 2nd ed. John Hopkins University Press, Baltimore, MD.

REGIONAL AND FIELD GUIDES

AFRICA

Eardley, C., M. Kuhlmann, and A. Pauly. 2010. "The Bee Genera and Subgenera of sub-Saharan Africa." *Abc Taxa* 7, 1–138.

Pauly, A., R. W. Brooks, L. A. Nilsson, Y. A. Pesenko, C. D. Eardley, M. Terzo, T. Griswold et al. "Hymenoptera Apoidea de Madagascar et los Iles Voisines." *Annales Sciences Zoologiques* 286. Musée Royal d'Afrique Centrale, Tervuren.

ASIA

Soh, Z.W.W. and J. S. Ascher. 2020. *A Guide to the Bees of Singapore.* National Parks Board, Singapore.

AUSTRALASIA

Donovan, B. J. 2007. "Apoidea (Insecta: Hymenoptera)." *Fauna of New Zealand* 57. Manaaki Whenua—Landcare Research New Zealand, Lincoln.

Houston, T. 2018. *A Guide to Native Bees of Australia.* CSIRO Publishing, Clayton.

Smith, T. J. 2018. *The Australian Bee Genera: An Annotated, User-Friendly Key.* University of New England Press, Armidale.

EUROPE

Else, G. R. and M. Edwards. 2018. *Handbook of the Bees of the British Isles.* 2 vols. The Ray Society, London.

Polaszek, A. 2011. *Identification Key to the European Genera of Bees (Insecta: Apoidea).* Natural History Museum, London.

NORTH AMERICA

Holm, H. N. 2017. *Bees: An Identification and Native Plant Forage Guide.* Pollination Press LLC, Minnetonka, MN.

Messinger Carril, O. and J. S. Wilson. 2021. *Common Bees of Eastern North America.* Princeton Field Guides. Princeton University Press, Princeton, NJ.

Michener, C. D., R. J. McGinley, and B. N. Danforth. 1994. *The Bee Genera of North and Central America.* Smithsonian Institution Press, Washington, DC.

Williams, P. H., R. W. Thorp, L. L. Richardson, and S. R. Colla. 2014. *Bumble Bees of North America: An Identification Guide.* Princeton Field Guides. Princeton University Press, Princeton, NJ.

Wilson, J. S. and O. Messinger Carril. 2015. *The Bees in Your Backyard: A Guide to North America's Bees.* Princeton Field Guides. Princeton University Press, Princeton, NJ.

Images of most of the bee species of Canada can be found at https://www.beesofcanada.com/species by Cory Sheffield and on the author's York University web page cited above.

SOUTH AMERICA

Silveira F. A., G.A.R. Melo, and E.A.B. Almeida. 2002. *Abelhas Brasileiras: Sistemática e Identificação.* Fundação Araucária, Belo Horizonte.

INDEX

All index entries relate to bees, unless otherwise specified (e.g., wasps).

Families, subfamilies, and tribes are in Roman type; genera and species are in *italics*.

The publisher would like to thank the following for permission to reproduce copyright material:

Adam J. Martinez, April 2017: 53. **Adriana Tiba and Julio Pupim**: 70, 186, 225.**Alamy Stock Photo** Anthony Bannister: 27; Antje Schulte—Insects: 169; Arya Satya: 228; BIOSPHOTO: 18BL, 49, 94, 128, 193, 220, 232; blickwinkel: 13, 16L, 34, 44, 51, 54, 78, 157, 163, 192, 206; blickwinkel, BIOSPHOTO: 85; Bob Gibbons: 142; Bryan Reynolds: 74; Charles Melton, 25L, 37, 69, 77, 144, 155, 160, 161, 170, 183, 196, 216, 217, 75; Daniel Borzynski: 106; David W. Williams: 28BR; Denis Crawford: 80, 89; Eduardo Estellez: 195T, FLPA: 17, 162; Francisco Lopez-Machado: 200; imageBROKER: 108; Janos Rautonen: 229; João Burini: 214; Joe Dlugo: 90, 91; John Cancalosi: 197; Minden Pictures: 98–9, 180, 201; Premium Stock Photography GmbH: 138; Richard Becker: 166; Science Photo Library: 194; WildPictures: 158. **Alexandre Callou Sampaio**, Barbalha-Ce, Brasil: 64–5. **Amro Zayed**: 31L, 31M, 31R, 66. **Annalie Melin**: 50. **Bernardo Segura**: 102. **Bernhard Jacobi**: 112, 124, 125, 230. **Bob Peterson** (CC BY-SA 2.0): 30B. **Bug Guide** Aaron Shusteff: 60; Jillian H. Cowles: 86. **Cheryl A. Fraser**: 12. **Chien Lee**: 223T. **Cristiano Menezes**: 30T, 213B. **DavidFrancis34** (CC by 2.0): 131. **Dino J. Martins**: 48. **Dr Kit Prendergast: the Bee Babette**: 110, 141, 179T, 179B. **Dreamstime** Amskad: 199; Edward Phillips: 176, 181; Hakoar: 174; Helen Ifill: 202; Henk Wallays: 58; Henrikhl: 47; Jean Landry: 101; Kirill Kolyshev: 6; Manfred Ruckszio: 40; Paul Reeves: 107; Sandra Standbridge: 35. **FAO/Nature Kenya/Dino Martins**: 84. **Francisco Santander**: 147. **Gail Fraser**: 118. **Getty Images** Arterra, 59; Auscape: 167; Bildverlag Bahnmuller: 38; Paul Starosta: 7; Scott Spakowski: 198; Valter Jacinto: 171. **Hans Bänziger**, 'The remarkable nest entrance of tear-drinking *Pariotrigona klossi* and other stingless bees nesting in limestone cavities (Hymenoptera: Apidae)', by H. Bänziger, S. Pumikong, and K. Srimuang, *Journal of the Kansas Entomological Society*, 84: 22 –35, 2011: 32. **Hans Bänziger**, *Vampire Moths: Behaviour, Ecology and Taxonomy of Blood-sucking Calyptra*, Natural History Publications, Kota Kinabalu, Malaysia, 2021: 209B. **Heather Holm**: 39, 190. **iNaturalistbee. iN** Vijay Anand Ismavel: 224. **Invasive.Org** Allan Smith-Pardo, Exotic Bee ID, USDA APHIS PPQ, via Bugwood.org ITP Node, 23. **iStockphoto** Heather Broccard-Bell: 139; scharag: 36; sherjaca: 178; Yves Dery: 95. **Jason Gibbs**: 87. **Jean and Fred Hort**: 133, 140. **Jenny Cullinan**: 130. **Jesús Moreno Navarro**: 222, 223B. **João Burini**: 135. **Judy Gallagher**: 218. **Kerry Stuart**: 24, 122, 231. **Kushal Kulkarni**, India: 209T. **Laurence Packer**: 26T. **Léo Correia da Rocha Filho**: 184. **Liam Graham**: 8, 9L, 56T, 148, 185, 219. **Linda Rogan**: 113. **Marama Hopkins**: 126. **Matías Gargiulo R**: 103. **Miranda Kersten** (CC BY-SA 2.0): 68. **Nature Picture Library** Andy Sands: 22L, 168; Kim Taylor: 28T; Phil Savoie: 22R. **NJ Vereecken**: 88, 154. **Pablo Vial**: 61, 62, 123, 134, 137, 146, 152, 187, 215. **Patrich Cerpa**: 120, 121, 226, 227. **Peter Vos**: 165B. **Pierre Bornand**: 188. **Quentin Vandemoortele**: 67. **Robert C. Turner** @rctphotos: 43. **Sebastián Andrade Trujillo**: 116, 191. **Shawn O'Donnell** (CC BY 4.0): 208. **Sheila Dumesh**: 11B, 56B. **Shutterstock** arthit vicharn: 204; Brian Lasenby: 97; Celso Margraf: 210; Dmytro Lopatenko: 173; Eataru Photographer: 233; Ed Phillips: 5, 43, 114; Eduardo Dzophoto: 221; Elliotte Rusty Harold: 207; Frank Reiser: 18BR; Gabriel Spenassatto: 213T; Glenn McCrea: 156; Henrik Larsson: 46, 79, 105; HWall: 10, 16R, 72, 104, 150, 172, 195B; Kamieniak Sebastian: 25R; Keith Hider: 109; Krasowit: 28BL; Kuttelvaserova Stuchelova: 42; LABETAA Andre: 177; lcrms: 41; Luc Pouliot: 100; Luisaazara: 211; Mathisa: 205; Murilo Mazzo: 203; Paul Reeves Photography: 96; Peter J. Traub: 129; Tacio Philip Sansonovski: 212; Wirestock Creators: 18T, 151. **Siegwalt**: 164–5. **Simon Oliver**:

92–3, 189. **Spencer K. Monckton** 2013: 119. **Steve Buchmann**: 26B. **Tâmara Miranda Lomba**: 65B. **Thomas Onuferko**: 182. **Tim Rudman**: 127, 132. **Tom Astle**: 83. **USGS Bee Inventory and Monitoring Lab** Sam Droege, 2, 9R, 52, 159, 136, 149. **Wikimedia Commons** Gideon Pisanty (CC BY 3.0) (https://commons.wikimedia.org/wiki/File:Ochreriades_fasciatus_male_3.jpg): 153; KRHick (CC BY-SA 4.0) (https://commons.wikimedia.org/wiki/File:Perdita_minima_.jpg): 76. **Wikipedia** Israel Gideon (CC BY-SA 3.0) (https://commons.wikimedia.org/wiki/File:Camptopoeum_frontale_male_1.jpg): 71; Israel Gideon (CC BY-SA 3.0) (https://commons.wikimedia.org/wiki/File:Melitturga_male_2.jpg): 73.

All reasonable efforts have been made to trace copyright holders and to obtain their permission for the use of copyright material. The publisher apologizes for any errors or omissions and will gratefully incorporate any corrections in future reprints if notified.

ACKNOWLEDGMENTS

I am most grateful to the photographers, listed above, who allowed me to use their images in this book and to the production team listed on the inside cover for putting up with me. I thank Susi Bailey for diligent wordsmithing and Caroline Elliker for her work getting this book started.

For helpful discussion, comments and/or information related to this book I thank John Ascher, Hans Bänziger, Cecily and Robert Bradshaw, Matthew Clark, Sheila Colla, Esteban d'Bzi, Michael Dillon, Sam Droege, Sheila Dumesh, Connal Eardley, Cheryl Fraser, Felipe Freitas, Liam Graham, Marama Hopkins, Terry Houston, Michael Kuhlmann, Annalie Melin, Cristiano Menezes, Spencer Monckton, Thomas Onuferko, Michael Orr, Christophe Praz, Kit Prendergast, Sandra Rehan, Fernando Silveira, Jeff Skevington, Simon Tierney, Pablo Vial, George Walker, Ken Walker, Paul Williams, H. M. Yeshwanth, and especially Stephen Buchmann, Jack Neff, and Olivia Messinger Carril.

I thank my past students who have enriched my understanding of bees immensely.

Without the love and support of Gail Fraser, I would not be able to accomplish much, and I thank her from the top, sides, center, and bottom of my heart.

This book is dedicated to my granddaughter Lynn.

ABOUT THE AUTHOR

Laurence Packer is a distinguished research professor in melittology at York University in Toronto, Canada, where he has worked since 1988 and teaches courses mostly in entomology and biodiversity. He and his students have published more than 230 research articles on bees and described over 200 species and two new genera. Laurence oversees one of the most diverse bee collections in the world, with 90 percent of the bee genera represented and with bees from more than a hundred countries. In addition, he cochairs the campaign to obtain DNA barcodes for the bee species of the world, more than 25 percent of which have now been sequenced.

Laurence is the author of *Keeping the Bees* (2010) and coauthor, with Sam Droege, of *Bees: An Up-Close Look at Pollinators* (2015). He is also an editor for the journal *Insect Conservation and Diversity*. Current projects include a revision of the species of the genus *Xeromelissa* (which will involve describing more than 40 new species), as well as a revision of the apoid wasp genus *Parapiagetia* from South America.